SCIENCE 101

OCEAN SCIENCE

HarperCollins books may be purchased for educational, business, or sales promotional use. For information, please write: Special Markets Department, HarperCollins Publishers, 10 East 53rd Street, New York, NY 10022.

Produced for HarperCollins by:

Hydra Publishing
129 Main Street
Irvington, NY 10533
www.hylaspublishing.com

FIRST EDITION

Library of Congress Cataloging-in-Publication Data

Hoffman, Jennifer.
  Science 101 : ocean science / Jennifer Hoffman. – 1st ed.
      p. cm.
  Includes bibliographical references and index.
  ISBN: 978-0-06-089139-8
  ISBN-10: 0-06-089139-4
  1. Oceanography–Popular works. I. Title. II. Series.

GC21.H585 2007
551.46–dc22

                                                    2007060872

12   13   14   SCP   10 9   8   7   6   5   4   3   2

SCIENCE 101

OCEAN SCIENCE

Jennifer Hoffman

Collins
An Imprint of HarperCollinsPublishers

CONTENTS

WELCOME TO OCEAN SCIENCE

Left: Mushroom soft coral, featured in the deep-sea collection at the Monterey Bay Aquarium. Its tentacles contain stinging poisonous cells that capture plankton drifting past. Top: A lush coral reef, with hard and soft corals, anthias, and golden damselfish. Amazing collections of biodiversity, coral reefs are among the most ecologically threatened places on Earth. Bottom: A Caribbean reef squid cruising through black water. Small and social, the Caribbean reef squid lives in schools of four to thirty squid. It is semelparous, meaning that it dies after reproducing.

On a boat off the coast of Alaska, an oceanographer notices that the water is a milky turquoise rather than its usual dark color. She looks at a water sample through a microscope and finds that, as she suspected, the cause of the color change is the presence of coccolithophores, a type of single-celled alga encased in hard, round plates. She cannot tell the size of the planktonic aggregation, or bloom, from her vantage point, but images from a specially designed satellite orbiting hundreds of miles above her show that this bloom extends for thousands of square miles. Before 1997, such blooms were rare, but they have occurred in most years since then. What is causing this change in the waters of the Bering Sea?

This story illustrates the breadth of scale that exists within the field of oceanography. Our seas are studied from space, through microscopes, and from in the midst of them. By observing, measuring, and modeling, scientists generate and answer a multitude of questions using a variety of tools as they piece together the story of the oceans. And since most of the Earth is ocean, the story of the oceans is the story of Earth.

EXPLORING THE MYSTERIES

In the words of eminent marine biologist Rachel Carson, "We can only sense that in the deep and turbulent recesses of the sea are hidden mysteries far greater than any we have solved." Scientists who study the seas are drawn to these mysteries, to the practical and intellectual challenges involved in deciphering such a foreign realm. Areas of investigation cover the past, present, and future, as well as the full range of scientific disciplines—physics, chemistry, geology, and biology.

Over the centuries, oceanographic studies have revolutionized scientific understanding of the world. Patterns of magnetic stripes on the seafloor and the contours of the bottom of the ocean give clues to the mechanisms behind the slow movement of continents. The remains of tiny organisms in marine sediments dating back millions of years contain the story of mass extinctions, and of sudden and lethal drops in oxygen levels in the deep sea. The vibrant biological communities around hydrothermal vents miles below the surface prove that life can sustain itself without energy from the Sun. These and other discoveries provide vital clues about how the world works.

Scientists are mostly motivated by the thrill of discovery, but there are many practical benefits to studying the marine realm. Powerful anti-cancer compounds have been found in marine animals, and scientists are exploring commercial applications of the sticky substance mussels use to attach themselves to rocks. Coastal wetlands can be used as natural water treatment plants, and healthy mangrove forests help protect coastal communities from flooding, storms, and tsunamis. Predicting the trajectory of climate change requires understanding what drives ocean currents and the interaction between ocean and atmosphere, as does predicting whether hurricanes will get stronger and more frequent. Even the microscopic organisms in the sea play a key role in regulating climate by taking up carbon, which limits the buildup of greenhouse gases in the atmosphere.

Saltwater mussels in an intertidal zone along the Alaskan coastline.

STUDYING THE SEAS

Off the coast of Florida, sitting in an acrylic sphere just five feet in diameter, a scientist hundreds of feet below the surface of the ocean studies the flashes of light produced by a multitude of strange creatures. Along the coast in California, students carefully count and catalog the species there, looking at how Earth's changing climate has affected the composition of marine communities. Oceanographers bounce sound waves off the seafloor to investigate underwater landslides that may have caused catastrophic tsunamis around Hawaii. A remotely operated vehicle sent to the bottom of the ocean to look for hydrothermal vents finds instead a volcanic eruption in progress and returns to the surface covered in golden droplets of elemental sulfur.

Marine habitats range from the highly variable world at the ocean's edge to the stable fields of mud on the abyssal plains to the blue waters of the open sea. These varied habitats and the organisms that exist in them require different approaches, with each approach providing a different piece to the puzzle of the world's oceans. *Science 101: Ocean Science* provides just a glimpse of the diversity of puzzle pieces that scientists have gathered, and opens the door to further exploration.

A red sea star clinging to coral. Sometimes called "starfish," sea stars are not fish but echinoderms, belonging to the same phylum as sea urchins and crinoids. Fossils of sea stars have been dated to the Ordovician period, which began roughly 488 million years ago.

A Steller's sea lion cow cuddling with her pup on Lowrie Island in Forrester National Wildlife Refuge, Alaska. Scientists have classified over 100 living species of marine mammals.

An endangered loggerhead turtle. All species of sea turtles are threatened or endangered. Sea turtles can live for up to 189 years. They are acutely sensitive to Earth's magnetic field and, upon maturity, return every two to four years to nest on the same beach from which they hatched.

3

CHAPTER 1

THE VAST UNKNOWN

Left: People have long been enthralled with the immensity of the world's oceans. Today there is still much to be explored under the waves. Top: The Earth viewed from space. Oceans cover almost three-quarters of Earth's surface. Bottom: People have long relied on the ocean for commerce; today, trips around the world for trade are commonplace.

It is unlikely that an airplane equipped with the most up-to-date maps could crash into an undiscovered mountain. How could a mountain thousands of feet tall be missing from a map? Yet on January 7, 2005, a mere 400 miles (644 km) from its base on Guam, a U.S. Navy submarine ran into an uncharted seamount. According to official charts, the sub was cruising through water around 6,000 feet (1,829 m) deep. One crew member died and 23 others were injured.

Although the ocean covers almost three-quarters of Earth's surface and makes up more than 90 percent of the habitable area, or biosphere, surprisingly little is known about it. The surfaces of the Moon and Jupiter have been mapped more thoroughly than the floor of the sea has.

Understanding the ocean matters. It matters because the ocean plays a major role in determining climate and weather. It matters because we use the oceans day in and day out for transportation, food, medicine, and pleasure. It matters because billions are spent exploring space while our own backyard is relatively unexplored. And it matters because what we learn by studying the oceans is just so fascinating.

One Ocean, Many Regions

Looked at from the shore, the ocean may appear to stretch unchanging to the horizon. Except for those bodies of water surrounded by land, all the world's oceans are connected. Yet just as there are distinct regions of the surface world—rain forests, tundra, and prairies, for example—there are distinct regions within the ocean. And just like terrestrial organisms, some ocean creatures travel the world, while others are restricted to very particular conditions. To understand and protect the vast diversity of life in the ocean, we need to understand and protect the vast diversity of habitats and processes that support it.

LOCATION, LOCATION, LOCATION

The ocean can be divided into two basic regions: the pelagic zone, or water column, and the benthic zone, or seafloor. There are further divisions within those two zones based on depth and proximity to shore. The pelagic zone is divided into coastal (neritic) zones and oceanic zones (those away from the influence of land), and into the mesopelagic, bathypelagic, and abyssal pelagic zones, in order of increasing depth.

The divisions of the benthic zone begin high on shore in the supralittoral, or splash, zone, which is rarely if ever covered with water. Then there is the intertidal zone, regularly submerged and exposed as the tides rise and fall, followed by the sublittoral, or subtidal zone, which extends to the edge of the continental shelf. The deepest zones are the abyssal and the hadal. The hadal zone encompasses the very deepest areas, in the trenches or anywhere below 19,800 feet (6,000 m).

These place-based divisions reflect physical differences, such as the amount of freshwater input or annual changes in temperature, that have a profound influence on the biology and chemistry of each underwater region.

INVISIBLE BOUNDARIES

From the perspective of organisms that need light to make energy (that is, plants and other photosynthesizing organisms), there are two distinct zones in the ocean. One is the photic, or sunlit, zone, and the other is the aphotic, or lightless zone. Yet another division reflects changing water chemistry and density. Water that is warmer or

Above: The seafloor, or benthic zone, extends from the splash zone to the very deepest areas of the ocean. Many types of plant and animal life inhabit all of these areas. Top left: The seashore lies in the coastal, or neritic, zone of the ocean.

less salty is less dense than cooler or saltier water, and water that is less dense tends to stay at the surface. Between the surface and deep zones is the pycnocline, or zone of rapidly changing density. Although invisible to us, the pycnocline is a very real boundary for the small creatures that make up the majority of life in the oceans.

Currents form another set of boundaries in the ocean. The water is full of planktonic organisms, or organisms that are weak swimmers. Traveling against or even crossing major currents can be impossible for these beings.

DISTINCT COMMUNITIES

Ocean zones can also be categorized based on the dominant features of particular communities. On land, there are forests of trees; in the oceans, there are kelp forests. On land, there are tall-grass prairies; in the oceans, there are sea-grass meadows. Some communities, such as mangrove forests and salt marshes, straddle the area between land and sea. Other examples of ocean communities include coral reefs, mussel reefs, hydrothermal vents, and methane seeps. Each type of community is distinct, offering a unique set of opportunities and challenges for the organisms that live there.

Top right: These pelagic deep-water shrimp were collected at a depth of more than 3,000 feet (914 m) in the Canada Basin. Bottom right: Bull kelp grows very quickly. The seaweed can grow from a tiny spore into a 200-foot-long (60 m) plant in just months.

The Teeming Hordes

Envision a world in which some animals as well as plants remain fixed to the ground, getting food by grabbing passersby or filtering it from the air around them. Imagine that not just the ground but the air itself was full of life—microscopic organisms invisible to the naked eye but whose presence colored the air, innumerable jellyfish, odd-looking snails with wings. Imagine that at night, the vast majority of these animals could produce their own light, flashing signals to one another through the dark. This is the world of the oceans. They are full of life—some of it familiar, some of it bizarre, and much of it still unknown.

ANIMAL DIVERSITY

Taxonomists divide animals into about 33 major groups, called phyla. Each phylum represents a distinct body plan. The difference between phyla is much greater than the difference between species within an individual phylum. Out of all these phyla, only one has no representatives in the ocean: the little velvet worms called onychphorans. By contrast, 14 phyla are found only in the ocean. The large-scale diversity of animals in the oceans far outshines that on land. Even though the number of identified species on land is much higher than that in the oceans, this number is deceptive, as new ocean organisms continue to be discovered. For example, in 1988, scientists realized that what had been viewed as just two species of commercially valuable deep-sea crabs was really

Above: Coral reefs are among the most diverse ecosystems on Earth. The range of life forms in this habitat is sometimes compared with that in tropical rain forests. Top left: Bell jellies have long, thin tentacles that are used to capture crustaceans.

Hydrothermal vents on the ocean floor provide a habitat for organisms that have adapted to this extreme environment.

The Mir *submersible is a battery-powered vessel that can carry three people to a depth of 20,000 feet (6,000 m).*

CENSUS OF MARINE LIFE

In its own words, the Census of Marine Life (CoML) is "a growing global network of researchers in more than 70 nations engaged in a ten-year initiative to assess and explain the diversity, distribution, and abundance of marine life in the oceans—past, present, and future." This ambitious project grew out of the concern that diversity in the oceans was disappearing before scientists were able to research it. CoML activities are amazingly varied. Some researchers use electronic tags to track the movement of large predators such as sharks or whales. Others create computer models to predict the effects of fishing or pollutants. Still others work to figure out just how the abundance of various species has changed by sequencing DNA or even by reading through hundred-year-old restaurant menus that show a connection between meal prices and relative abundance of many fish species.

18 different species. Every year brings similar findings—in 2001, a 23-foot (7 m) squid unlike any known species was discovered; in 2003, it was Big Red, a bloodred, two- to three-foot-long (60 to 90 cm) jellyfish with no tentacles; 2004 brought a hot dog–sized new species of dragonfish, discovered on a trip to the relatively well-studied Bear Seamount off New England.

THE UNDERAPPRECIATED MULTITUDES

Although animals receive more attention, they are a minority when it comes to ocean life (and life on Earth, too). The seas are packed with plants, algae, bacteria, and protists. The diversity is stunning, although a microscope is required to appreciate much of it.

Microscopic primary producers, which are single-celled algae and other microscopic creatures capable of making their own energy, are the foundation of most marine food webs (just 5 to 10 percent of marine productivity comes from large algae and plants). These tiny organisms are able to make their own energy from sunlight or chemicals.

On hydrothermal vents, bacteria use either sulfurous compounds or light from the vents to make energy. In the open ocean, bacteria as well as dinoflagellates, diatoms, and a host of other phytoplankton make energy from sunlight. Dinoflagellates have also gotten a lot of attention for their ability to produce nasty toxins. In large numbers, they cause so-called red tides, which can be fatal for a range of marine life.

Mother Ocean

Although we talk of Mother Earth, the term "Mother Ocean" might be more appropriate. Life almost certainly began in the oceans, and most major groups of animals had their beginnings there. It was marine microbes that produced the oxygen that makes the atmosphere hospitable to humans and gave rise to the ozone layer that shields Earth's inhabitants from the Sun's damaging ultraviolet radiation. The world's oceans currently generate almost three-quarters of the atmosphere's oxygen and hold 97 percent of Earth's water. And the oceans harbor more different phyla of animals than does any other habitat.

Why is it that the ocean houses such a great diversity of life relative to land and freshwater? Part of the answer may simply be that there is so much more of it. From the surface to the bottom, the ocean provides usable living space.

It is also possible that since life began in the oceans, organisms have had more opportunity to diversify there. It also may be easier for organisms to stay put than to move into a completely new environment. But there are physical and chemical reasons that the oceans are such good places to live as well.

WET AND SALTY

One key advantage that water has over land or air is that water is always wet. Organisms living in the water do not have to develop special ways to keep critical membranes moist. While mammals have their lungs

Above: Sandpipers taking flight in New Brunswick, Canada. These shorebirds can be found on coasts and in wetlands around the globe. Top left: This shell-less pelagic snail uses both its muscular body and a pair of "wings" to hunt other pelagic snails. These and other minute sea creatures are collectively known as plankton.

Much of the coloration along this shoreline comes from algae. These chlorophyll-containing organisms have been around for more than two billion years and can take in water anywhere along their surface.

tucked inside their bodies where evaporation is minimal, many marine animals have their breathing apparatuses on the outside, waving about in the current. While land plants have to pull moisture up from their roots, algae can take up water anywhere along their surface. Also, water can hold many nutrients in dissolved form, providing nourishment for algae, plants, and animals.

While the vast majority of terrestrial and freshwater organisms spend a good deal of energy regulating the salt concentration of their internal fluids, the blood and body fluids of most marine organisms have the same salt content as the surrounding water. (Key exceptions to this rule are most fish and all mammals.) Another effect of saltiness is that seawater is denser than freshwater, which makes floating in the ocean easier than in freshwater bodies, like lakes and ponds.

ADRIFT AND TASTY

The ocean's wetness, saltiness, and nutrient content have fostered the evolution of a vast floating community, the plankton. Throughout the ocean, bacteria, protists, and many algae and animals spend their lives drifting with the currents. Many animals that normally live on the seafloor cast their young adrift for days, months, or even years. The presence of so much life in the water has fueled the evolution of a feeding mechanism not seen on land: suspension feeding. Some suspension feeders hold out their arms, tentacles, or other feeding appendages and wait for food to run into them. Others suck water across a filter or mucous net, concentrating particulate matter. Suspension feeders may float through the water among food items, or they may sit on the ocean bottom.

The gills of a whale shark are vertical slits located in front of the pectoral fin. Sharks, like most fish, rely on their gills to filter oxygen from the water.

A Variety of Perspectives

The ocean is a complex place, and understanding it can prove to be a complex business. Oceanography, the study of the oceans, is divided into four primary disciplines—chemical, physical, geological, and biological. Even so, many questions require an integrated view, and subdisciplines such as geochemistry or even biogeochemistry have emerged over time.

STUDYING LIFE

People who study life in the oceans are often divided into marine biologists, who focus on life close to shore, and biological oceanographers, who study life farther out at sea. Both groups study individual behavior, population cycles, interactions among multiple species, and the interactions between living organisms and their nonliving environment. Life in the ocean is so different from life on land that studying marine organisms can radically shift a scientist's perceptions of how life works. For instance, the discovery of deep-sea hydrothermal vents proved that not all ecosystems depend on sunlight as their ultimate source of energy. Tracking global patterns of primary productivity with satellite maps helps scientists to understand marine communities, currents, and climate.

Above: A scientist gathers samples at the shore. Top left: A scuba diver collects sponges and sea squirts for research.

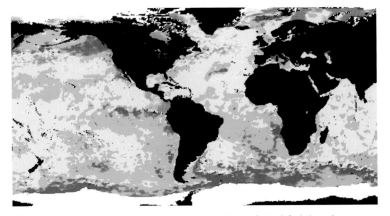

This chart from the U.S. National Oceanic and Atmospheric Administration (NOAA) shows a global analysis of sea surface temperature. The dark blue areas are the coldest waters; light blue, yellow, and orange represent progressively warmer temperatures. The white areas indicate sea ice.

STUDYING EARTH STRUCTURES

Geological oceanographers look at sediments, rocks, fossils, and the topography of the seafloor. They may also study coastal processes such as erosion and deposition of new sediments. These scientists may look through microscopes to identify minerals or microscopic fossils, or use satellites to track the movement of sediments beneath the surface. The theory of plate tectonics, which is less than 50 years old, has radically changed the perception of land and sea. The theory explains the long mountain ranges found under the sea, the patterns of earthquakes and volcanoes, and why the seafloor is so young.

STUDYING MATTER AND ENERGY

Physical oceanographers focus on physical processes in and affected by the oceans. They study worldwide patterns of circulation, temperature, winds, and salinity. They look at how light and sound travel through the water and examine the physical interactions of ocean and air. They study how waves form and move across the ocean. In addition to researching large-scale phenomena, physical oceanographers also observe small-scale ones. They research questions such as how animals use microscopic cilia to generate feeding currents, or how a sea urchin larva's shape affects how it moves through the water. Ocean physics can be beautiful and elegant, yet eminently practical.

A typical sea ice station showing an ice corer, a generator for power, sleds for hauling gear, and light and temperature sensors.

STORIES FROM THE ICE

Polar ice is like a history book: neat stacks of dust, microorganisms, air pockets, and ice in chronological order. Reading this book requires researchers from many disciplines. The Greenland Ice Core Project (GRIP) ice core was drilled 9,930 feet (3,029 m) deep and goes back 200,000 years. Layers of dust mark major volcanic eruptions, including those of Italy's Mount Vesuvius in 6955 BCE and North America's Mount Mazama in 5676 BCE. Analysis of heavy metals tells of extensive atmospheric lead pollution between 2,500 and 1,700 years ago, reflecting mining and smelting by Greeks and Romans.

The longest, deepest ice core was drilled in Antarctica. It is 10,728 feet (3,270 m) deep and holds information from as long as 650,000 years ago. The chemical composition of the ice gives scientists the average global temperature throughout the time period, and air bubbles trapped in the ice give samples of the ancient atmosphere. From the data gathered at the site, researchers concluded that not only do carbon dioxide levels rise and fall with average temperature but that they are higher now than they were at any time in the past 650,000 years.

STUDYING ATOMS AND MOLECULES

Some would claim that, in the end, everything comes down to chemistry. And the ocean is a rich place indeed to study chemistry. Almost every element in the periodic table can be found in the sea. Chemical oceanographers study the chemical composition of seawater and how it influences the quantity and types of life that occur in different areas of the ocean. They track the fate and distribution of chemical pollutants and the ways in which organisms use chemical compounds to protect themselves. These scientists measure the movement of carbon dioxide between the sea and the atmosphere, a process that is critical to understanding and predicting changes in the global climate.

Intertwined Futures

Humans depend on the ocean for sustenance, commerce, and inspiration. The ocean is also responsible for pulling literally tons of carbon dioxide out of the atmosphere every year, helping to regulate climate change. It moves heat around the globe, keeping western Europe much warmer than it would otherwise be. Chemical compounds produced by marine organisms are turning out to be useful in fighting cancer. The future of humans and that of the ocean are intimately connected.

LEAVING OUR LEGACY

People marvel at the spectacular beauty of coral reefs. It is difficult to imagine anyone sitting by while boats dragged huge nets across these reefs, leaving a barren sea floor. Yet it happens. Coral reefs in the deep sea are destroyed at an alarming pace as deep-sea trawlers essentially bulldoze the bottom as they dig for shrimp and fish. These corals, which live as deep as 3.7 miles (6,000 m), in cold, dark environments, are no less spectacular than those that live in shallow waters. Deep-sea explorers have only recently discovered the extent of these coral gardens. An expanse 37 square miles (100 sq km) in size was discovered off the coast of Norway in 2002.

Around the world, humans are radically altering marine ecosystems. Through overharvesting, pollution, and habitat destruction, the populations of many species have been reduced to the point of possible extinction. Climate change adds to these stresses on the environment, making recovery from or adjustment to other stresses even more difficult.

Local, regional, and international organizations have sprung up to protect the oceans and their inhabitants. Stretches of ocean are being set aside as protected areas. Restaurants, stores, and individual consumers are using sustainability as a criterion for the fish they buy and sell. And thousands of volunteers work to monitor coral reefs, protect sea turtle nesting grounds, and educate the public about marine ecosystems.

In the grand sweep of geological time, extinctions are not unusual. At several points throughout Earth's history, well over half of existing species became extinct. Each extinction event occurred over the course of a million years or so, a relatively short period of

Above and top left: The beauty of life under the sea should never be taken for granted. The human impact on ocean life can be devastating, but in many countries people are taking action to preserve underwater habitats.

A female leatherback turtle digs into a sandy beach on the African coast to lay her eggs. Of the eight species of sea turtles in the world, all six that are found in U.S. waters are endangered.

time from a geologic point of view. The diminishing of biodiversity is an aesthetic as well as a practical loss. Each species is a work of art. Some may be more beautiful than others, but once lost, none of them can be replaced.

NEW FRONTIERS

Recent innovations in ocean science make this an exciting field of study. Inventions such as submersibles, underwater robots, and a range of remote sensing technologies have opened up new views of the ocean, allowing for a truly global look at marine processes. Yet there is still much more to discover. Once unimaginable ideas—for example, a vast network of automated sam-

pling and monitoring stations thousands of feet below the surface—may soon become reality. In the next few decades, great strides will be made toward a better understanding of currents, tectonic plates, diversity, and even simply what the bottom of the ocean looks like. Many more "monsters" of the deep will surely be discovered. Though it has been going on for thousands of years, the exploration of the oceans has only just scratched the surface.

A Conductivity-Temperature-Depth (CTD) instrument is lowered into the water in the Gulf of Mexico. These instruments help scientists investigate patterns of salinity, temperature, and density in the ocean.

CHAPTER 2

THE HISTORY OF OCEANS

Left: Scientists believe that the Earth was formed 4.5 billion years ago. The Hadean period, named for the Greek underworld, Hades, was characterized by extreme heat and volcanic activity. Top: An artist's representation of early Earth. Bottom: Humans have a long history of ocean exploration. The USS Vincennes, *commissioned in 1826, was the first U.S. Navy ship to circumnavigate the globe. The* Vincennes *continued her world travels for nearly 40 years.*

The ocean is old, but not quite as old as Earth itself. The seas arose from a violent world of volcanic eruptions and asteroid impacts, followed by 20 million years of heavy rain, to become the probable birthplace of life. Over time, life diversified and radically altered the land, air, and sea. Periods of mass extinction have several times shifted the balance of power among living beings.

For thousands of years, humans have looked to the ocean for food, resources, transportation, and inspiration. They have sought to understand disasters such as tsunamis and hurricanes. Many have simply been motivated by curiosity about the world under the waves.

Traversing the sea in boats of all kinds, humans learned to navigate using the stars, Sun, and Moon. They also discovered how to read currents, waves, and wind for faster and safer travel. Early maps showed what different cultures knew about an area, and over time, maps have become more complex with increased knowledge of land and sea. The history of oceanography is the history of humans answering the ocean's call.

Ocean Beginnings

Scientists believe that about 4.5 billion years ago, a cloud of gas, dust, and debris swirled around the Sun. Slowly, material clumped together into small planetesimals, which collided and fused and gradually created the planets. Collisions and near collisions between growing planets and other orbiting bodies generated tremendous amounts of heat. The fiery nature of this period in Earth's history inspired its name: the Hadean period, named after the Greek underworld, Hades.

As the Earth cooled, heavier materials were pulled inward by gravity to form its iron-rich core, while lighter elements became the mantle and crust. Gases were also released, creating an atmosphere. As cooling progressed, water condensed in the atmosphere and fell as rain. Initially, this rain probably evaporated back into the atmosphere almost as quickly as it fell, but as the Earth continued to cool, liquid water began to accumulate. A period of heavy rains that lasted some 20 million years formed the oceans. Sedimentary rock was formed around 3.9 billion years ago, as the weight of water compressed mud and sediment over time. A new period, called the Archean, began. Earth was probably one vast ocean with scattered islands, although the very earliest continental plates may have formed during this time. It was in the Archean period that life began.

The next phase in Earth's history, the Proterozoic, was also a time of great change. Lasting from between 2.5 billion and 543 million years ago, it saw the creation of the first real continents, the accumulation of oxygen in the atmosphere, and the appearance of the first multicellular life.

EARLY LIFE

It is likely that life originated in the oceans. After all, there was virtually no land when life first appeared. But in what part of the ocean did life begin? Charles Darwin envisaged "some warm little pond, with all sorts of ammonia and phosphoric salts, light, heat, electricity, etc." But with no ozone layer, the surface of the Earth would have been bombarded with levels of ultraviolet (UV) radiation lethal to most life forms. Based on the knowledge that light is not essential for life, many researchers favor the idea of life beginning in environments much like today's hydrothermal vents, or at least water that was deep enough to avoid deadly levels of UV radiation.

The oldest fossils are around 3.5 billion years old, created by cyanobacteria, or blue-green bacteria. These bacteria are capable of photosynthesis and can form mounds of layered bacteria and calcium carbonate called stromatolites. Living stromatolites exist today; the best known examples are in Australia's Shark Bay.

Above: An artist's depiction of a Mosasaurus, *a large, carnivorous marine lizard that grew to 29 feet (9 m) in length. Top left: Soon after the formation of Earth, a fragile crust was formed that was breached by meteorite impacts and volcanoes.*

The early atmosphere contained very little free oxygen, but bacteria did not need oxygen to live. In fact, bacteria produced oxygen as a waste product. By two billion years ago, this waste product began to build up in the atmosphere. Many bacteria probably died off in this toxic new atmosphere. Others found sheltered places with no oxygen. New life forms arose, with organisms capable of using oxygen rather than being poisoned by it. The oxygen also helped build the ozone layer, giving life in the shallow seas some protection from the Sun's harmful ultraviolet rays. The first organisms with a cell nucleus appear in the fossil record some 2.1 billion years ago. By 1.3 billion years ago, multicellular algae appeared, and finally, 600 million years ago, the first multicellular animals appeared.

Right: This diorama, which is displayed at the National Museum of Natural History in Washington, D.C., illustrates how Earth may have looked in the Cretaceous period (144 to 65 million years ago). Top: These stromatolite reefs in Shark Bay, Australia, may be thousands of years old.

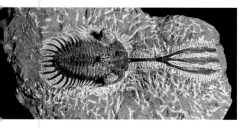

Mass Extinction

As organisms evolve and diversify, it is inevitable that not all species will survive. Indeed, most species that have ever lived are now extinct. Sometimes their population gets so small that finding a mate is almost impossible. Sometimes another species outcompetes them for food or other resources. These are background extinctions, and, like the death of an individual, they are a normal part of life.

Every so often, the pattern of extinction changes. Instead of a single species here or there, huge numbers of species die off at once. In the history of Earth, there have been five or six times when more than half of all species disappeared at the same time. Such events are called mass extinctions.

MOTHER OF ALL EXTINCTIONS

The best-known mass extinction on Earth is the end-Cretaceous, or K-T, extinction, which occurred 65 million years ago. Yet the Permo-Triassic extinction 245 million years ago was far more devastating. In the span of less than a million years, up to 95 percent of all marine species were lost, along with 80 percent of amphibian species and 90 percent of reptilian species. In the oceans, one of the most striking losses was that of the trilobites, which had been a major part of marine communities for more than 300 million years.

At a distance of more than 200 million years, it is hard to be sure what caused this "great dying." One contributing factor may have been the significant drop in sea level during the Permian period. This would have destroyed vast amounts of marine habitat, changed the pattern of ocean currents, and altered where and how much primary productivity, or photosynthesis, occurred. Another cause may have been the widespread occurrence of hypoxic areas, zones with so little oxygen that most animals could not survive in them. Yet a third suspect is volcanic activity. Within just a million years, between half a million and three-quarters of a million cubic miles (2 to 3 million cubic kilometers) of lava poured across Siberia. This much volcanism would have put enough dust and debris into the atmosphere to block much of the incoming light, cooling the Earth and reducing primary productivity. Another theory is that a sizable asteroid or comet struck Earth.

REBIRTH

One interesting aspect of mass extinctions is that there seems to be little relationship between which groups succumb to a mass extinction and which die off during normal times. In

Left: The chalk cliffs on the coast of southern England date back to the Cretaceous period. Top left: A trilobite fossil. These animals inhabited the seas 540 million years ago.

A fossil ammonite. Although they survived the biggest mass extinction of all at the end of the Permian, these nautiluslike mollusks disappeared completely during the K-T extinction.

other words, the organisms that die as a result of catastrophe are not just the marginal ones. Animals that had been dominant have been known to disappear completely within the geological blink of an eye (a million years or so). The rapid loss of so much diversity, while disastrous, creates a new beginning. Each mass extinction event was followed, eventually, by a burst of diversification as surviving groups reproduced and began to fill the many empty niches left by the extinction.

Just as some lineages are lost forever in mass extinctions, the complete reshaping of the ecological and evolutionary landscape allows others to rise and flourish. Members of our own class, the mammals, were mostly furtive, rodentlike creatures until the almost complete annihilation of the dinosaurs. Mammals are now diverse and widespread and include the largest species on the planet.

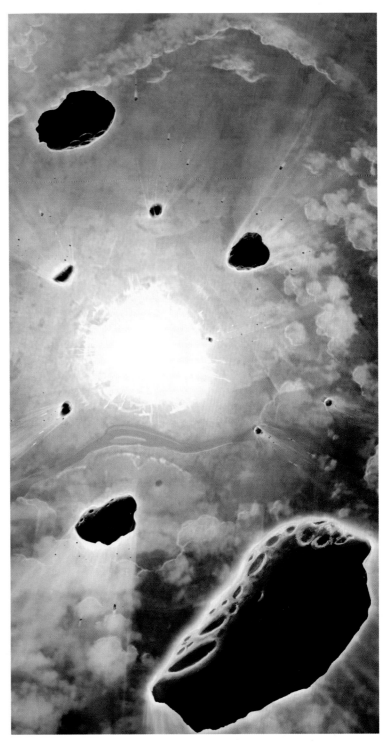

The K-T extinction, which led to the death of the dinosaurs, may have been caused by an asteroid striking the Earth.

Early Explorers

Above: England's Francis Drake was the first captain to sail his own ship around the world. Top left: The peoples of the Pacific Islands were among the earliest to travel long distances on the ocean.

While *Homo sapiens* began as land-based hunter-gatherers, humans have long been intrigued by the sea. Perhaps the earliest and most wide-ranging of the ocean voyagers were the Polynesians. Their journey probably began near Borneo between 4,000 and 6,000 years ago and took them throughout the 10 million square miles (26 million sq km) and 10,000 islands of the central and South Pacific. The Hawaiian Islands, more than 2,000 miles (3,200 km) from the closest significant Polynesian island, were last to be settled, but travel between these new islands and those to the south became common. How did Polynesians manage to navigate across the open ocean, far from any land? They were exceptional navigators, using not just stars, but the patterns of winds, waves, and currents to guide them.

AROUND THE MEDITERRANEAN

On the other side of the world, many Mediterranean cultures made good use of their enclosed sea for commerce, sustenance, and empire building. The earliest may have been the Minoans, a Bronze Age culture based in Crete, but other cultures were not far behind. Temple reliefs in Egypt depict an expedition led by Queen Hatshepsut in 1478 BCE. Five 70-foot-long ships with sails and oars set out through the Red Sea, returning with frankincense, myrrh, and other valuables and curiosities.

The Phoenicians are often considered the greatest of early Mediterranean mariners. They dominated trade on the Mediterranean Sea for centuries, reaching the peak of their civilization between 1000 and 500 BCE. Trading ideas as well as goods, Phoenicians probably brought knowledge that sparked the Golden Age of Greece. In fact, the word *bible* comes from the Greek name for one of the great Phoenician ports, Byblos. Phoenicians may even have been the first to circumnavigate Africa. According to the ancient Greek historian Herodotus, the Phoenician navigator Hanno completed this trip around 500 BCE. Hanno's statement that the noontime Sun shone in the north at the southern tip of Africa, in direct contrast to its location in the Mediterranean, gives proof of this journey.

BEYOND THE MEDITERRANEAN

While others headed south, the Vikings explored widely in the North Atlantic around 1005 CE. They even built a settlement in North America, but it lasted just a few years. It appears that Vikings continued crossing the Atlantic regularly for more than 300 years, bringing wood back to their settlements in Greenland.

Asians and Arabs took to the sea as well. Classic sea tales such as "Sinbad the Sailor" often contain accurate information about sea routes and culture. The quadrant, an early navigational instrument, may trace its roots to a ninth-century Arab invention. From the other side of Africa, Portugal sought a sea route to India, succeeding in 1498.

Spain looked westward, sponsoring Columbus's trips to South America and the Caribbean in the 1490s and

Far ahead of European mariners, the Chinese built ships with multiple masts that could carry hundreds of passengers. During his stay at China's imperial court from 1275 to 1292, Marco Polo described four-masted, seagoing merchant ships with watertight bulkheads and crews of up to 300. China's fifteenth-century ships traveled as far as Africa to trade goods.

TREASURE SHIPS

Many key maritime inventions trace back to the Chinese: The rudder, the compass, and the presence of multiple watertight compartments in the hold are just a few. The matt-and-batten sails characteristic of Chinese sailing ships had several advantages over other sail types. They allowed ships to sail almost directly into the wind, remained functional even when torn, and could be furled and unfurled from the deck, meaning no one had to climb up masts in high winds. China had extensive maritime contacts and in the early fifteenth century sent out the famed treasure ships, named for the gifts they carried to countries ranging as far west as modern Kenya. Of course, the ships also collected tributes to bring back to China. By the late fifteenth century, China had shifted to an isolationist policy: The construction of seagoing ships was prohibited, and China's days as a maritime power ended.

Ferdinand Magellan attempted to sail around the world for Spain. He died during the voyage in 1521.

Magellan's attempted round-the-world voyage. Although Magellan was killed before completing the expedition, others of his group became the first to circumnavigate the globe, finishing their voyage in 1522. Magellan gave the Pacific Ocean its European name; after the rough and stormy waters at the south end of South America, he was impressed by the seeming calm of the "new" ocean.

England's Sir Francis Drake was the first to successfully lead an entire expedition around the world, from 1577 to 1580. Although motivated by trade more than by science, these voyages set the stage for the scientific explorations that followed.

The Power of Maps

Knowledge is power, and a map, at its most basic, is a way of depicting knowledge that has to do with place.

The oldest known map, a nine-foot-long wall painting in Turkey that dates to 6200 BCE, shows the plan of the town in which the painting was found. Another map, this one engraved on a palm-sized piece of baked clay in 2500 BCE by an unknown Babylonian, shows an area with mountains and a river and identifies the owner of the land. What people put on a map tells what matters most to them. Cultures for which the coasts and ocean were important paid just as much attention to mapping the sea as they did to mapping the land. The Inuits of the far north have for centuries produced stunningly accurate maps of their homeland's coast-

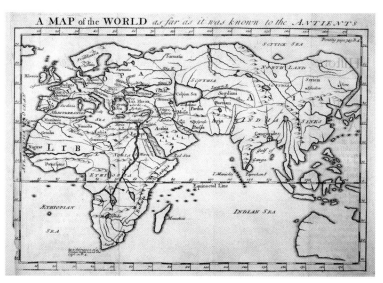

Above: This map is a re-creation of a fifth century BCE "map of the world" by Herodotus. The map ends just beyond the shoreline. Although Herodotus doubtless left out many details in order to get the "whole world" on a single map, it is clear the seas mattered much less to him than to the makers of the stick chart (below). Top left: In the American colonies, cattle horns were etched with maps as decorative pieces.

line, despite the absence of any special navigational equipment. The early Polynesians produced maps from sticks and shells. The shells were islands, and the sticks depicted the patterns of swells, in some cases perhaps of currents or winds. Thus they had "landmarks" in the waves of the ocean. It is likely that these early

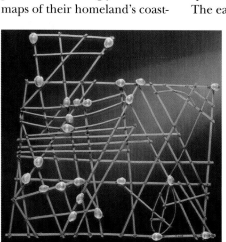

A Polynesian stick chart. Each shell represents an island, and the sticks show the orientation of swells.

mapmakers did not use the maps themselves but created them to communicate with others.

MEDITERRANEAN RIVALRIES

The Mediterranean Sea has been surrounded by many countries for thousands of years. Not surprisingly, this situation has led to frequent rivalries over trade routes, and to the use of the sea to wage war and expand empires. Maps could be the key to success or failure in such campaigns. The Phoenicians, for instance, kept detailed maps,

The library of Alexandria, established over 2,000 years ago, was to contain one copy of every book ever written, and had copies of every map ever brought to its port. It housed books by Plato, Aristotle, and Socrates, among others.

updating them whenever they discovered new dangers or routes that were safer or faster than existing ones. Yet this information was shared only with Phoenicians or friendly nations. A story tells of a Phoenician captain who ran his ship onto the rocks rather than let a Roman ship that was following him see how to safely access the rocky shores of the Isles of Scilly.

Such rivalries continued for centuries. A Portuguese map made in 1630 was referred to as the "Secret maps of the Americas and the Indies from the Portuguese archives." During the same time period, maps in Spain were often locked in royal vaults and shown only to trusted officials.

ALEXANDRIAN SCHOLARSHIP

After the founding of the great library of Alexandria around 300 BCE, all boats visiting the port of Alexandria were searched for books and maps. These items were copied by scribes, the originals kept by the library, and the copies returned to the ships' captains.

Egypt was a fertile land for ideas during this time period. The Greek scholar Eratosthenes came up with the first accurate estimate of the circumference of Earth. He used simple algebra, the distance between Alexandria and Aswan, and differences in the angle of the Sun at these two locations. Close to 100 years later, another Greek revised Eratosthenes's correct estimate downward by a quarter, an error that stuck until the fifteenth century CE.

Time and Space

For maps to be useful, users need to know where they are when they read them. On or near land, this is not difficult, since visual references such as headlands or rivers are plentiful. But what if there are no such landmarks? One solution is to create a grid of intersecting lines and determine location relative to *x* and *y* coordinates. This approach, using latitude and longitude, was suggested as early as the second century BCE by the Greek mathematician Hipparchus and put into widespread use by Ptolemy in 150 CE.

The grid is made of up lines that circle the Earth from pole to pole (longitude) and lines that run parallel to the equator (latitude). Latitude describes north-south position; longitude

A seventeenth-century navigator uses a device to locate his position using the stars. Top left: A map of the world as Ptolemy envisaged it in 150 CE.

Above: A brass strip marks zero longitude at Greenwich, England. Great Britain was the first to set a standard time throughout a region when it established Greenwich Mean Time (GMT) in the 1840s. By 1855, most public clocks in Britain were set to GMT. Other countries followed, and in 1884 the International Meridian Conference met in Washington, D.C., and established international time zones.

describes east-west. The position of the Sun, Moon, or stars provides enough information for an individual to calculate latitude. Calculating longitude, however, is more difficult. While latitude is absolute—it is calculated relative to the equator—longitude is arbitrary. The line indicating zero degrees longitude could in theory be placed anywhere, and in fact was located in a variety of different spots before it was finally, definitively placed in Greenwich, England. In 1884, 25 countries agreed to use the Greenwich meridian as the prime meridian, and over the next several years more and more countries adopted it as the

standard. Over the ages, various people suggested solutions for calculating longitude at sea, but none worked with any degree of reliability and accuracy. The problem was grouped with those of finding the fountain of youth or a way to turn lead into gold.

THE CHALLENGE OF TIME

In 1530, the Dutchman Regnier Gemma Frisius suggested using clocks to measure longitude. By carrying a clock set to the time at a known longitude and comparing that with local time, which could be determined by a sighting of the Sun at its zenith, you could tell how far you had traveled in an east-west direction. Since Earth completes a rotation every 24 hours, each hour it travels one-twenty-fourth of that distance, or 15 degrees of longitude. Therefore, if you could accurately determine the time difference between noon at your location at sea and noon at a given location ashore—Greenwich, England, for example—you could determine your longitude.

Unfortunately, Frisius's idea predated sufficiently accurate seagoing clocks by more than 200 years. By the mid-1600s, pendulum clocks were capable of measuring time accurately on land, but the rolling motion of a ship at sea made them completely unreliable for marine travel.

As ocean travel became more important, and European countries fought over who owned which of the newly discovered islands, several countries offered increasingly greater rewards to anyone who could devise a method of accurately measuring longitude at sea. In 1714, four homeward-bound British warships miscalculated their longitude and ran aground, losing 2,000 men. In response, Queen Anne authorized the creation of England's Board of Longitude and offered the biggest prize yet: Anyone with a method to determine position to within half a degree of longitude would win 20,000 pounds, the equivalent of several million dollars today. Proposals poured in, and by the end of the century the so-called longitude problem had been solved.

British inventor John Harrison (1693–1776). Harrison, a watchmaker, was the first person to accurately calculate longitude aboard a ship. This was an important development in global navigation.

A WINNING CLOCK

In 1735, an English watchmaker named John Harrison, with no formal education, presented a pendulum-free clock to the Board of Longitude. It was virtually friction free, would not rust, and was unaffected by the rolling movement of sea travel. After Harrison and his clock performed better than a ship's navigator on a voyage between Lisbon and London, the Board of Longitude had a hard time refusing Harrison's claim to the prize. But refuse it they did. Forty years, several new clocks, a prize from the Royal Society, and the intervention of King George passed before Harrison finally got the prize he had so ably earned. Why the difficulty? The Board of Longitude, under the influence of Sir Isaac Newton, saw longitude as a problem for astronomers, not for a common watchmaker.

The Dawn of Scientific Exploration

James Cook's 1772 trip on the HMS *Resolution* marked a turning point in ocean voyages. It was the first to use Harrison's new chronometer to calculate longitude and thus the first to create an accurate map of the Pacific.

The voyage of the English ship the *Beagle* from 1831 to 1836 was significant for another reason. While its purpose was to map regions of the South American coast, Charles Darwin was also a passenger on this journey. It was during his travels on the *Beagle* that Darwin formulated his theory of evolution by natural selection. This theory, along with the theory of plate tectonics, revolutionized the way scientists saw the world.

The late eighteenth to early nineteenth century saw the rise of scientific analyses of the sea. Matthew Fontaine Maury, an officer in the U.S. Navy, used wind and current logs to generate maps of global patterns of these phenomena. Maury went on to publish what many consider to be the first oceanography textbook, *The Physical Geography of the Sea*, in 1855. He also organized the first international meteorological conference, which globalized the study and collection of weather data at sea.

OCEANOGRAPHIC RESEARCH BEGINS

One of the most influential oceanographic voyages was

Left: A replica of the HMS Endeavor. *The ship was used by Captain Cook on his first voyage. Cook carried no chronometer on this trip. Top left: The HMS* Challenger *under sail. Above: Charles Darwin formulated his theory of natural selection while on the five-year surveying journey of the HMS* Beagle.

Britain's 1872 *Challenger* expedition, whose goal was to investigate "everything about the sea." The *Challenger* was the first ship specifically outfitted as an oceanographic research vessel, with laboratories for biological, physical, and chemical oceanography. It circumnavigated the globe, traveling more than 62,000 miles (100,000 km) and visiting all the major oceans except the Arctic. Over the course of the journey, the *Challenger* made some 362 sampling and observation stops. At each station, circumstances permitting, the crew collected the same information in the same way: They measured depth, temperatures throughout the water column, speed and direction of currents, and meteorological conditions. They sampled the seafloor and bottom water and used trawls or dredges to study whatever life was on the bottom. They towed nets through the water at different depths to sample life in the mid-water and at the surface. When the *Challenger* returned, more than 100 scientists helped to examine the samples. The 50-volume *Challenger* report is a classic document in oceanographic science.

Institutions dedicated to studying the oceans sprang up around the world. In 1888, the Marine Biological Laboratory in Woods Hole, Massachusetts, was founded, as was a center for marine biology in Plymouth, England. These institutions continue to be among the most influential institutions of marine biology today.

A rainbow off the coast of Maui. The Hawaiian Islands were settled by Polynesian explorers sometime between 340 and 600 CE.

MAUI'S MAGICAL HOOK

A Polynesian demigod named Maui is a common figure in legends across the Pacific and is credited with the formation of islands ranging from Hawaii to New Zealand. Although there are some variations from culture to culture, the basic story is this: The young Maui, told by his brothers that he could not go fishing with them, hid in their boat. When the boat was far out at sea, Maui revealed himself and cast his magic fishhook into the sea. As Maui and his brothers pulled in his line, they pulled land out of the sea. In the Maori version of the story, Maui worried that the gods would be angry and left his brothers with the fish while he went to pacify the gods. His brothers argued while he was gone, and their clubs created the mountains and valleys of New Zealand's North Island.

THE ICY NORTH

One reason the *Challenger* expedition skipped the Arctic Ocean was the pack ice. Getting trapped in that ice spelled almost certain doom for a ship, as the force of the ice would crush it. Yet Fridtjof Nansen, a Norwegian, was curious about currents in the Arctic and decided to build a ship that could withstand the Arctic conditions.

Nansen and shipbuilder Colin Archer designed a boat, the *Fram*, that was not only incredibly strong, but also squat and rounded so that when caught in the ice it would be pushed up rather than crushed. In late September 1893, Nansen and his crew set out in the *Fram* with the goal of getting caught in pack ice off Siberia. They would drift with the ice for several years until, they hoped, they reached the western edge of the Arctic Ocean. The plan worked beautifully, and after three years the *Fram* came out as watertight as when she had started. Nansen, who left the ship in a failed bid for the North Pole in March 1895, returned separately.

Modern Exploration

From the invention of echo sounding and scuba diving to the development of ocean-specific satellite programs, the past hundred years have seen an explosion in oceanographic technology. In no small part because of this expanded ability to "see" into the depths, understanding of the ocean and its inhabitants has blossomed as well. Since the International Decade of Ocean Exploration (1971–80), long-term international research programs such as the Global Ocean Observing System have helped scientists understand the ocean on a truly global scale.

SEEING THE DEPTHS

The greatest difficulty in mapping the seafloor is that it is impossible to see from the surface. Early oceanographers used the line-and-sinker method, in which a weighted line was slowly lowered to the bottom and raised back up. The length of rope that was let out gave a rough measurement of the depth. Along with its tendency to overestimate depth as currents pulled the line off perpendicular, this method had two key flaws: the sheer volume of rope needed and the time it took (often all day) to get just a single depth measurement. In the late eighteen hundreds, scientists came up with the idea to use sound instead. The amount of time it took for a noise generated at the surface to echo back from the bottom would show how far away the bottom was. In a weeklong trip across the Atlantic in 1922, the first successful deep-ocean echo sounder made three times as many soundings as the *Challenger* had in her three-and-a-half-year voyage. The *Meteor*, a German ship, used an echo sounder to make the first detailed map of a section of seafloor as it crisscrossed the Atlantic between 1925 and 1927.

Although an immeasurable improvement on older methods, echo sounding still could measure depth only at a single location. This limitation was overcome by sonar.

WAR AND OCEAN EXPLORATION

After the devastating use of submarines by the Germans in World War I, British scientists developed a way to locate submarines underwater. Called sonar

Above: A diagram of the so-called wire-drag system for surveying the seafloor. Prior to the development of side-scan sonar, this was the only reliable method of finding protrusions from the bottom when surveying the seafloor. Top left: The bathyscaphe Trieste II.

(sound navigation and ranging), it used sound waves to detect the size, speed, and location of underwater objects. It works because of the way sound travels in water. Sonar was widely used in World War II, but it was not until the 1960s that its potential for seafloor mapping was realized.

Inspired by a system used by astronomers, a group of U.S. Navy and civilian scientists expanded traditional sonar into multibeam sonar in 1962. Using multiple transmitters and receivers, they were able to measure the depth along a swath of seafloor as wide as the water was deep. A similar technology, side-scan sonar, uses differences in echo intensity to generate actual images of the bottom.

THE VIEW FROM SPACE

Although space might seem an odd place from which to study the oceans, satellite information has expanded our understanding of oceans tremendously. There are satellites that measure sea surface temperature and the amount of phytoplankton on the surface, both of which give information about currents. Over the past three decades, global geodetic satellites have evolved into a powerful source of data for studies of the Earth, its oceans, atmospheric systems, and interactions between them. The U.S. Navy commissioned the Geodetic Satellite (GEOSAT) in 1985. This satellite is capable of measuring the height of the ocean's surface to within two inches (5 cm)—not bad, from 830 miles (1,336 km) up! Because sea surface height reflects bottom topography, *Geosat* has allowed oceanographers to create a rough worldwide map of the seafloor. Anything smaller than 6 miles (10 km) across can easily be missed by this method, so researchers continue work to develop the technology.

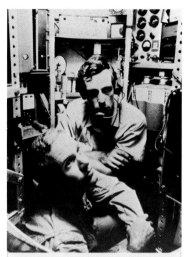

Lieutenant Don Walsh and Jacques Piccard were photographed inside the bathyscaphe Trieste *in 1960 at the Mariana Trench.*

SEEING THE BOTTOM

The first trip to the deep sea was in 1934, when William Beebe and Otis Barton descended to 3,072 feet (923 m) in the bathysphere, a 4,500-pound steel sphere tethered to a ship. They never saw the ocean bottom. But in 1960, a new submersible, the *Trieste*, carried two people to the bottom of the Mariana Trench, 35,800 feet (10,916 m) below the surface. No submersible since has gone as deep.

With several manned submersibles currently in use, the number of people who have seen the bottom of the deep sea is steadily growing, as is our appreciation of this bizarre region. The recent discovery of deep-sea coral reefs indicates how much of the seafloor remains to be explored.

A NASA Satellite over Dungunab Bay, Sudan. Satellites like these can provide valuable information on regional or global oceanographic processes.

CHAPTER 3

EVER-CHANGING EARTH

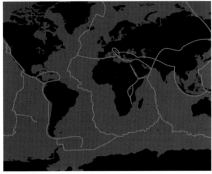

Left: Coral reefs, home to numerous marine species, are susceptible to changes in climate, salinity, and sea level. Top: Bust of Aristotle, 384–322 BCE. Bottom: Tectonic plates, outlined in red. As these plates shift, they can move continents, build mountains and trenches, and cause earthquakes and volcanic eruptions.

Aristotle wrote that "a place does not always remain land or sea throughout all time, but where there was dry land there comes to be sea, and where there is now sea, there one day comes to be dry land… But the whole vital process of the Earth takes place so gradually and in periods of time which are so immense compared with the length of our life, that these changes are not observed."

Aristotle was about 2,000 years ahead of his time. His ideas anticipated today's knowledge of geological processes. For centuries the Christians who dominated intellectual thought in Europe held that such ideas were blasphemous, and scholars who suggested the existence of an evolving Earth were silenced. Yet rumblings of dissent were inspired by the study of fossils and landforms. This research matched features between continents that are now far apart. Later investigations into the bottom of the ocean further supported theories of moving continents. Scientists began to see that the Earth might not be as solid as once thought. Rather, continents and the plates on which they sit make slow but steady progress, floating along on a sea of semisolid material. The average continent moves at about the same rate at which human fingernails grow.

Layer-Cake Earth

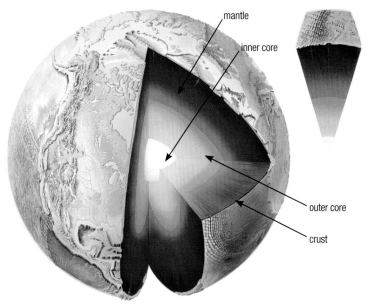

Mapping Earth's interior takes ingenuity. Scientists have used meteorites and lava for clues about what is below the surface, but the most useful tool has been seismography, the study of how waves travel through Earth. Just as a bell vibrates when it is struck, Earth vibrates after earthquakes, volcanic eruptions, and explosions. Since the chemistry and density of the material through which waves pass affect their speed and direction, the movement of waves can reveal a lot about where they have been.

Above: Earth's structure. The cutaway diagram shows the layer of upper rocky crust over the mantle of near-molten rock, shown in red. The molten outer part of the core is shown in yellow, the solid inner core in white. Top left: The crater of Mount St. Helens volcano, Washington State, which erupted in 1980.

WAVES

Two types of seismic waves travel through Earth. Primary, or P, waves travel fastest and are first to appear on a seismograph (a machine that records seismic activity) following an earthquake. They can pass through solids, liquids, and gases. Called compressional waves, they move by the rapid compression and expansion of material, the way a slinky moves when you push and pull it.

Secondary, or S, waves arrive later. Unlike P waves, they move from side to side, in an S shape, as would a rope if you laid it on a countertop and shook it from side to side. The inability of secondary waves to travel through liquid is key to mapping Earth's interior.

THE LAYERS

As Earth formed, different materials sank deeper or moved upward, depending on their density, in a phenomenon similar to oil floating on water. Just as the boundary between oil and water in a jar is very sharp, the shift between layers of Earth with different densities is also abrupt.

Three layers are clearly distinguished by density and chemical composition: the crust, the mantle, and the core. The crust is the least dense,

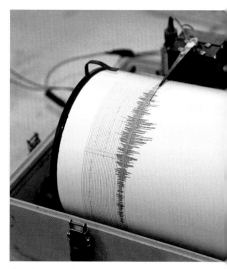

A seismograph recording ground movements, or seismic activity.

An iceberg is less dense than the water in which it floats, illustrating the principle of buoyancy.

although basaltic oceanic crust is denser than granitic continental crust. The mantle is more dense. At 1,781 miles (2,866 km) thick, it consists of silicon and oxygen, with lesser amounts of other materials, and can reach a temperature of more than 5,800°F (3,200°C). The core is even denser and hotter: It is around 90 percent iron and between 7,200°F and 9,900°F (4,000°C and 5,500°C). The inner core may be even hotter than the surface of the Sun!

Layers of Earth are also distinguished by how they behave. The crust and the uppermost mantle are like solids. Together they form the lithosphere. This rigid layer sits on top of the asthenosphere, a partially melted layer that behaves like a thick liquid.

The lowest portion of the mantle, though hotter than the

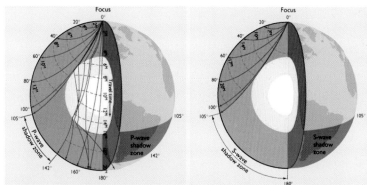

Diagram showing S and P wave "shadow zones." The diagram on the right illustrates the S wave shadow zone; the one on the left shows the P wave shadow zone.

SHADOW ZONES

When Richard Oldham identified S and P waves at the start of the twentieth century, he noticed something curious. As the waves passed through Earth, they traveled faster than expected through some regions and slower than expected through others. Further, no S waves ever made it to the region of Earth directly opposite an earthquake, and there was also a band where no P waves could be detected. Oldham realized that these S and P wave "shadow zones" meant the Earth must have a liquid core. A liquid core would block S waves completely and change the angle and velocity of P waves. Using more sensitive seismographs in 1935, Danish scientist Inge Lehmann noticed that very faint, low-frequency P waves actually sped up a little as they passed through the very center of Earth. Lehmann's observations led her to the conclusion that while Earth's outer core is liquid, there is also an inner, solid core.

asthenosphere, is under greater pressure and thus is less fluid. The core likewise has an outer, more liquid, portion and an inner, solid portion at the center.

ICEBERGS MADE OF ROCK

Because the asthenosphere acts like a liquid, buoyancy rather than material strength holds up the lithosphere. Massive as they are, continents are less dense than the layers below, just as an iceberg is less dense than the water in which it floats. The greater the density difference, the higher an object floats; an empty ship sits higher than a full one.

As with icebergs, much of a continent is beneath the surface, in the asthenosphere. Just how deep these continental roots reach changes over time. When a new mountain is formed, the crust around it sinks to accommodate the new weight. As mountains erode, they float higher. Melting glaciers have a similar effect. Areas that were covered by huge glaciers during the last ice age, 10,000 years ago, are still in the slow process of rebounding, floating higher. All of this rising and sinking of land creates local changes in sea level.

Continental Drift

Look at a map of the world and you will notice that many continents appear as though they could fit together like pieces of a jigsaw puzzle, particularly the east coast of South America and the west coast of Africa. This observation has fascinated scientists and others since the first good maps of both sides of the Atlantic began appearing in the sixteenth century. In the nineteenth century, the Austrian Eduard Suess, working in Germany, and Antonio Snider-Pellegrini, in France, added to the puzzle by pointing out the similarity between fossil organisms on the two sides of the Atlantic. Suess claimed that millions of years ago, the continents had been united in a single landmass, called Gondwanaland. He and others suggested that the oceans were formed when parts of the original continent sank or when land was eroded away during the biblical flood.

MOVING CONTINENTS

In 1915, German meteorologist and explorer Alfred Wegener published a comprehensive collection of data supporting ancient connections among continents. He presented information on similarities between fossils found on continents

Above: Artwork showing Pangaea, the supercontinent formed by continental drift in the late Paleozoic era (about 300 million years ago). Top left: This computer-generated image of the seafloor off the coast of Louisiana shows the continental shelf (red, orange, and yellow) and slope (light blue), and the abyssal plains (dark blue).

separated by oceans, mountain ranges and erosional features that matched up across the North Atlantic, and distinctive rock strata that were identical in South Africa and Brazil. Other examples were fossils of tropical plants and animals in distinctly nontropical locations, and coal found by famed explorer Ernest Shackleton in Antarctica. Wegener felt that these facts demonstrated not only that the continents had been connected, but also that they had moved. This idea was instantly ridiculed, in no small part because the mechanism suggested by

Wegener, centrifugal force, was quite inadequate. Yet the theory of continental drift refused to die. In 1935, Kiyoo Wadati of Japan suggested that the high number of earthquakes and volcanoes around Japan might be linked to moving continents. In 1940, Hugo Benioff plotted the location of deep earthquakes around the globe and found that in many cases they occurred along distinct lines that paralleled the continents. Many fell along the Mid-Atlantic Ridge, which had been discovered by the German *Meteor* expedition on its 1925–27 voyage.

Alfred Wegener (1880–1930), the German meteorologist who proposed the theory of continental drift.

This ridge also paralleled the continents. In the 1950s a boom in seafloor exploration yielded even more evidence.

The seafloor was found to be much younger than the continents, and its age increases the farther one moves from the mid-ocean ridges. Even more intriguingly, geophysicists discovered that either the North Pole or the continents must have moved significantly over the history of Earth. Iron-containing minerals act like compasses, pointing toward the North Pole, and in rocks of different ages they point in different directions.

A MECHANISM AT LAST

The next big step came in 1960 when Harry Hess at Princeton and Robert Dietz of Scripps Institute of Oceanography wrote papers putting forth the idea of seafloor spreading. They theorized that new seafloor is created at mid-ocean ridges as molten material from the mantle breaks through the crust. This results from giant circular convection cells in the mantle: Hot material rises, moves horizontally, cools, and sinks back down, only to be reheated and begin the process once again.

Continents, then, are passengers on much larger pieces of crust and lithosphere called tectonic plates. While new crust is created at mid-ocean ridges, or spreading centers, old crust is destroyed in areas called subduction zones, where one tectonic plate is pushed beneath another. The exact mechanism of plate motion is still debated in academic circles. It is likely that plate motion results from some combination of two mechanisms—the plates are pushed apart by the rising magma at spreading centers and pulled apart as the heavy leading edges sink into the asthenosphere.

Lake Powell, a man-made lake in Arizona. The geologic formations resulted from the advances and retreats of an inland sea that once covered the area.

CONTINENTAL DRIFT AND CLIMATE

As continents move around the surface of the globe, they affect more than just their own positions. Seaways open and close, isolating or connecting populations of marine organisms; land bridges are likewise created or destroyed. The position of the continents also affects global climate. During periods when much of Earth's landmass is closer to the poles, glaciers are more likely to form and spread. This process cools the global climate and causes sea level to fall. When sea level falls, shallow inland seas may disappear completely. There is evidence that evaporation of these seas can remove enough salt from the oceans to create a noticeable drop in salinity. Shallow ocean habitats are also lost when continents merge and inland climates become harsher. Continents that are joined together in one large landmass have more central area and less coast. Since being close to the ocean makes for a milder climate, continental interiors usually have more extreme climates than the coasts do.

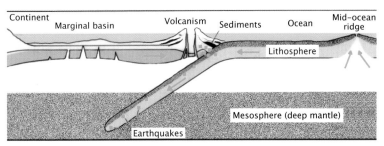

Diagram showing the formation of oceanic plates at a mid-ocean ridge and the subduction of an oceanic plate where it collides with a continental plate.

Where Two Plates Meet

The formulation of plate tectonic theory was an exciting event for geologists. After centuries of debate and frustration, there was finally a theory that could explain all the mysteries that had been laid out by Wegener and others, as well as the new observations being made about the ocean floor. It provided a new way to understand Earth and of course raised many questions. The most basic of these questions was, what exactly happens along plate boundaries?

PULLING APART

Where two plates are moving apart, a characteristic set of features forms. As hot material rises beneath the spreading center, the crust lifts and cracks, creating a so-called rift valley, with high ridges on either side. The ridges are often made up of a distinctive set of raised blocks, called horsts. As the plates move apart, magma rises from below, sometimes explosively in volcanoes, sometimes oozing. If the lava erupts slowly on a shallow underwater slope, it can form pillow lavas, so named because their rounded mounds look like heaps of pillows. Lava can also form into giant lakes. If these lakes drain before the lava hardens, strange pillars and "bathtub

Top: Seismic fault in Iceland, where the North American and Eurasian plates are diverging. Bottom: Pillow lava rocks off the coast of Hawaii. Top left: The Great Rift Valley, Kenya, Africa, created by the ongoing separation of the African and Arabian tectonic plates.

rings" may remain. There are even hot springs around spreading centers. First discovered in 1977, these hydrothermal vents support impressive communities of life. The mineral-rich chimneys can reach 180 feet (55 m) off the ocean floor.

Although spreading centers are usually hidden beneath the waves, the Mid-Atlantic Ridge makes a surface appearance around Iceland. Here the ridge spreads apart at a rate of about six inches (15 cm) per year.

COMING TOGETHER

Although divergent boundaries have tall ridges, they are nothing compared to the mountains that can form when two plates collide. One plate inevitably is pushed beneath the other, back into the asthenosphere, where it melts. As this plate melts, it creates a lot of magma and thus a lot of volcanic activity. This process creates chains of volcanic mountains, such as North America's Cascade Mountains, the Andes in South America, and the islands of Japan. Convergent boundaries build continents both through this volcanic activity and as a result of material being scraped off the descending plate.

Because continents are lighter than oceanic crust, when a continent meets an ocean, the oceanic crust is pushed beneath the continent. When two oceanic plates meet, the older, denser one is usually pushed beneath the other. Because both plates are dense, oceanic plate convergence zones can create very deep trenches, such as the Mariana

Trench, near Japan. The biggest mountains are created when two continents collide, because the continental crust is too light to be pushed down into the asthenosphere. The Himalayas, for instance, are the result of the Indian subcontinent plowing into the rest of Asia. They are still rising today.

SLIPPING SIDE BY SIDE

Convergent and divergent plate boundaries must bend to accommodate Earth's roughly spherical shape. This creates breaks in their otherwise linear nature, areas where plates are neither created nor destroyed, but simply slip by each other. The entire Mid-Atlantic Ridge is crisscrossed by a series of these so-called transform faults. Movement along transform faults is not always smooth. Sometimes the plates may move a bit before their sticky edges catch up with a sudden jerk. This motion can cause powerful earthquakes, most famously

The coastline of the Pacific Northwest. Volcanic activity created the Cascade Mountains, which are capped by snow and glaciers, visible here as white patches.

along the San Andreas Fault, in North America. Here the Pacific plate is grinding northwest at about two inches (5 cm) per year along its boundary with the North American plate. Unlike quakes in subduction zones, quakes along transform faults tend to be fairly shallow.

IS EARTH'S MAGNETIC FIELD WEAKENING?

Direct measurements of the strength of Earth's magnetic field go back as far as 1845. Since that time, the strength of Earth's magnetic field has decreased by close to 10 percent. We can use magnetic particles in ancient fired clay pots to gauge the strength of Earth's magnetic field even further back in time. English geologist John Shaw has found that the rate of decline in the last few centuries is higher than at any other time in the past 5,000 years. If this trend continues, the field could disappear altogether in 1,500 to 2,000 years, or perhaps reverse its orientation. On the other hand, we may simply be experiencing a magnetic lull. Still, the possible effects could be substantial. Not only do many animals rely on magnetic cues for migration, but the magnetic field shields Earth from significant amounts of incoming radiation. In areas where the field is weakest, satellites whose orbit is relatively close to Earth have already been damaged by such radiation.

Mapping Plate Boundarie

Knowing what happens at plate boundaries can also help to map them. As mentioned earlier, Hugo Benioff and others have mapped the location and depth of earthquakes around the globe; others have mapped the occurrence of volcanic eruptions. These maps form a striking pattern of interconnected lines across the surface of the globe. To confirm that these lines truly represent plate boundaries, more evidence is needed.

PALEOMAGNETISM

Where two plates move apart, new crust is formed on a regular basis as molten material rises from the mantle. As this magma cools, crystals form. The orientation of crystals containing iron or other magnetic elements is determined by the Earth's magnetic field. These crystals act like little compass needles, pointing north. Once the rock has solidified, the orientation of the crystals is fixed, so rocks reflect the direction of the magnetic field at the time they were formed.

In the late 1950s, geologists began to use a magnetometer to measure the magnetic field of rocks along the ocean floor. As they made more measurements, an odd pattern emerged.

There were distinct bands along the ocean floor, some of which had magnetic fields that enhanced Earth's, some of which subtracted from it. This meant that at regular intervals in Earth's past, the magnetic field was the opposite of what it is today. During those periods of reversed polarity, compasses would have pointed south rather than north. There have been 170 reversals over the past 76 million years, but such variability may not be the norm. During the Cretaceous period, Earth's magnetic field was stable for 40 million years, and during the Permian it was stable (although the opposite of today's) for 50 million years.

That Earth's magnetic field has reversed itself regularly is odd enough. Scientists were also mystified that the bands on either side of the Mid-Atlantic Ridge were mirror images of each other. In 1963, in a key development for plate tectonic theory, Drummond Matthews, Frederick Vine, and Lawrence Morley suggested that these stripes reflected formation of new seafloor at the ridge.

Magnetic seafloor stripes are also used to measure the rate at which plates are moving, and to match up areas of similar age around the globe.

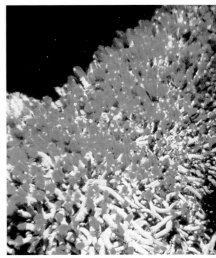

Top: Scuba diver with a magnetometer, used to measure the magnetic field of the ocean floor. Bottom: Tube worms on a "black smoker" hydrothermal vent on the mid-ocean ridge. Top left: The Himalayas reach an impressive 23,000 feet (7,000 m), but are dwarfed by Hawaii's Mauna Kea, which began as a seamount. Mauna Kea measures 33,000 feet (10,205 m) from base to peak.

HOT SPOTS

Not all volcanism happens on plate boundaries. The Hawaiian Islands are a classic example of hot-spot volcanoes, or areas with regular volcanic activity seemingly unrelated to plate boundary processes. These hot spots reflect plumes of superheated mantle rising up from Earth's core. For unknown reasons, hot-spot plumes remain relatively stable for millions of years. The Hawaiian hot spot has been active for at least 75 million years, during which time it gave rise to the Emperor Seamounts chain as well as the Hawaiian Islands. There is a sharp bend in the angle of the Hawaiian Island–Emperor Seamounts chain. Although the textbook explanation is that the Pacific plate changed direction while the hot spot remained stationary, current research suggests that the hot spot moved within the mantle.

Some mantle plumes are so large that they are referred to as superplumes. The biggest existing superplume stretches from Scotland to the Indian Ocean and from the Mid-Atlantic Ridge to the Red Sea. The action of this superplume is tearing the continent of Africa apart along its Great Rift Valley. It is also responsible for the formation of the Red Sea and the Gulf of Aden.

A "button" of uranium 235, which has a half-life of 704 million years, can be used for radiometric dating.

RADIOMETRIC DATING

Antoine Becquerel discovered the phenomenon of radioactive decay in 1896. Eleven years later, Bertram Boltwood demonstrated that the radioactive decay of uranium could be used to determine the age of rocks. Radiometric dating remains a critical tool in many scientific disciplines.

The basic idea is simple. Some elements have unstable nuclei whose decay (loss of protons, neutrons, or electrons) transforms one original element into another. For instance, uranium 238 decays into lead 206 (the numbers indicate the total number of neutrons and protons an atom contains), losing 16 protons and 16 neutrons in the process. Because this decay happens at a regular rate, it can be used to measure time. The standard unit is the half-life, the time it takes for half of the original element to decay. The half-life of uranium 238 is 4.5 billion years; that is, after 4.5 billion years, half of the uranium in a rock will have changed into lead 206.

Sounding the Deep

Ocean basins were once believed to be smooth, featureless plains across their entire area. Although most of the seafloor remains to be mapped in detail, we have enough information to understand the basic features of ocean basins. Our ability to "see" these features has played a critical role in developing the theory of plate tectonics and in understanding what happens where two or more plates meet.

RIDGES AND RISES

There is a remarkable system of ridges extending around the globe like seams on a soccer ball. These ridges, usually a mile and a quarter (2 km) or so in height, mark the zones where new ocean floor is being created. They may run down the center of an ocean, like the Mid-Atlantic Ridge. In this case, the ridge was formed when a continent split in half. In other cases, ridges may be closer to the edge than the middle. When plates move apart slowly, higher and steeper ridges result, like the Mid-Atlantic Ridge. When plates move more quickly, the ridges are more

Right: Bathymetric map showing one of several overlapping spreading centers on the East Pacific Rise. Top left: Inactive hydrothermal chimneys at Loihi Seamount.

gradual, as is the East Pacific Rise. These ridges and rises are also called spreading centers, because this is where two plates spread apart.

HILLS AND PLAINS

About a quarter of Earth's surface consists of abyssal plains, the true ocean floor. These featureless, sediment-covered regions are probably the flattest places on Earth. The sediment layer on these plains may be more than 3,200 feet (1,000 m) thick. Most of the sediment

comes from land rather than from activity in the ocean itself. Abyssal plains are less common in the Pacific, where most sediment eroded from the continent is trapped by trenches or island arcs rather than spreading out over the seafloor.

The plains are flat not because the underlying rock is flat, but because of the depth of sediment that has accumulated. The rugged contours created at spreading centers, or mid-ocean ridges, are still there but have been buried. Closer to the ridges,

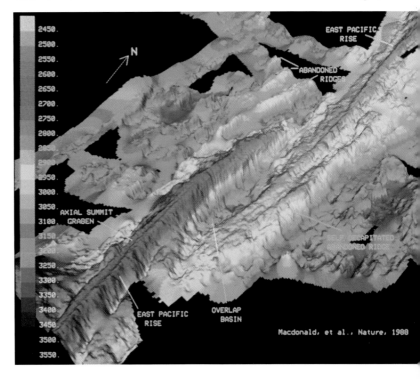

2450.
2500.
2550.
2600.
2650.
2700.
2750.
2800.
2950.
3000.
3050.
3100.
3250.
3300.
3350.
3400.
3450.
3500.
3550.

N

EAST PACIFIC RISE

ABANDONED RIDGES

AXIAL SUMMIT GRABEN

EAST PACIFIC RISE

OVERLAP BASIN

Macdonald, et al., Nature, 1988

there has been less time for sediment to accumulate, and the underlying hills and valleys start affecting the slope of the seafloor. Extinct volcanoes, hills, and mountains formed by the large volumes of molten lava arising at the ridge begin to show through. Because they are formed along spreading centers, abyssal hills usually occur in lines parallel to mid-ocean ridges and rises.

MOUNTAINS

Distinct from the abyssal hills are underwater volcanic mountains, called seamounts. They average close to 1,300 feet (400 m) in height, but the tallest seamounts rise more than 11,500 feet (3,500 m) above the ocean floor. They may form at hot spots, like those near the Hawaiian Islands, or near spreading centers. Some seamounts may become islands. Loihi Seamount, on the flank of Hawaii's Mauna Loa, is an active volcano and may one day break the surface to become an island. Other seamounts were once islands but are now completely submerged. This can happen because of rising sea level, erosion, or the sinking of a young volcano. These sunken islands are called guyots.

Although it seems implausible, big seamounts can be detected by increases in the height of the sea surface above them. Their gravitational pull attracts more water than the flat seafloor around them does. A 6,500-foot (2,000 m) volcano with a 12-mile (20 km) radius creates a 7-foot (2 m) increase in height of the sea surface.

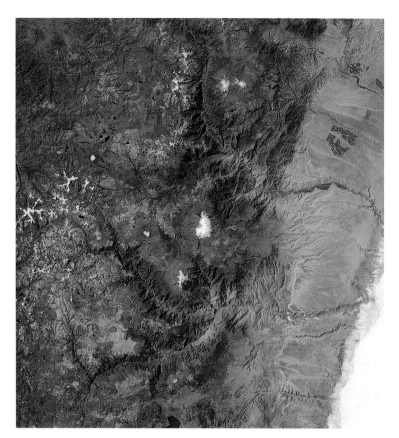

TRENCHES

Oceanic trenches are commonly found where oceanic plates converge and can dwarf the canyons found on land. Peru's Cotahuasi Canyon is the deepest land canyon at around 11,500 feet (3,500 m). In contrast, ocean trenches are generally 2 to 4 miles (3–6 km) deeper than the surrounding seafloor. The famous Grand Canyon does not get much deeper than 5,900 feet (1,800 m). Oceanic trenches are curved because of the spherical nature of Earth. They are steeper on one side than on the other. The slope on the seaward side, along the subducting plate, is more gradual.

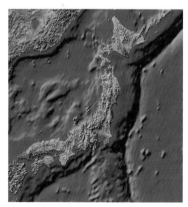

Top: Satellite photo of the world's two deepest continental canyons, Colca and Cotahuasi, in the Peruvian Andes. Bottom: This computer-generated image shows the Mariana Trench off the coast of Japan. The trench's deepest point is 6.7 miles (11 km) below sea level and more than 3.5 miles (6 km) below the nearby ocean floor.

Edge of the Abyss

For most people, the continents end where the ocean begins. For a geologist, continents can extend much farther than this. The lighter granitic crust characteristic of continents does not stop at the seashore, nor do continental processes such as erosion and sedimentation. The seafloor at the edge of a continent may have mineral deposits, sedimentary rocks, oil deposits, or other features that are characteristic of continents but not common to ocean basins. These underwater extensions, known as continental margins, may go on for hundreds of miles.

SHELVES, SLOPES, AND RISES

Closest to shore is the continental shelf, built to a large extent of material eroded from the main body of the continent. These sediments can be nine miles (15 km) thick or more. In areas with fast currents that carry sediments away, the shelf is thinner. In areas with extensive submerged reefs, the sediment layer is thicker. The shape, size, and topography of continental shelves are also determined by the degree of volcanic and tectonic activity in the area. On active continental margins, where a continent sits

atop a subduction zone, shelves are rarely very wide. The slope of the Andes, for instance, continues pretty much straight down in many places.

The true edges of the continents are the shelf breaks, which are the transitions from continental shelf to continental slope. Although not all that steep—the average is 4 degrees, and the maximum recorded is 25 degrees—the slope is noticeably steeper than the shelf. At the shelf break, lighter, thicker continental crust gives way to the denser, thinner oceanic crust. At the base of some continental slopes there

Above: This model of the Atlantic continental shelf and slope, with very exaggerated vertical relief, resembles an underwater Grand Canyon. The pagodalike spirals in the foreground represent seamounts rising from the ocean floor. Top left: The Grand Canyon.

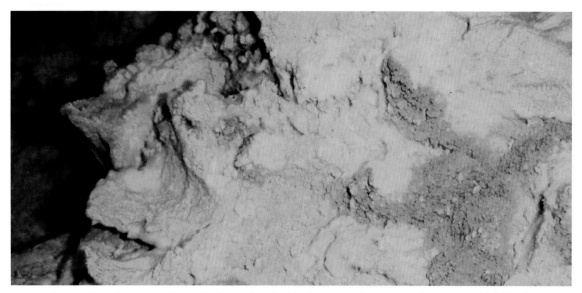

Lava rock covered in yellow iron oxide on the flank of the Loihi submarine volcano. Located southeast of the Island of Hawaii, the summit of Loihi is approximately 1 mile (1.61 km) below the ocean's surface.

may be one final feature before the true ocean basin begins—an apron of sediments called the continental rise. The biggest rises are associated with large rivers and are probably massive deposits of sediment tumbling down from the coast. In areas with strong currents or very steep topography, rises cannot form.

GRAND CANYONS

Impressive canyons often traverse continental shelves and slopes. Because these canyons often occur offshore of river mouths, geologists once thought they might represent erosion that had occurred during times of lower sea level. Yet while sea level has probably never been more than 600 feet (200 m) lower than it is today, many canyons extend beyond 10,000 feet (3,000 m) below current sea level.

The key to figuring out how these canyons formed came

from a 1929 earthquake off the coast of Newfoundland. After the quake, many transatlantic communication cables failed. They did not all fail at once, however. The cables closest to the epicenter failed almost immediately, and gradually cables farther and farther south of the epicenter failed. Cables as far as 300 miles (480 km) away from the earthquake failed. What happened? Scientists believe that there must have been some sort of underwa-

ter avalanche. As sediments mix with water immediately above them, they create a layer that is more fluid than sediment but denser than water. This layer will flow rapidly downhill in a turbidity current, reaching speeds of up to 17 miles per hour (27 km/hr). It may be that turbidity currents create submarine canyons. Canyons are self-perpetuating: Once formed, they become the easiest path for future turbidity currents to follow.

GEOLOGY AND POLITICS

How far away from its coastline can a nation claim control over marine resources? In 1945, U.S. president Harry Truman unilaterally declared that the United States controlled all natural resources from its coastline to the edge of its continental shelf. Other nations followed suit, and the United Nations Convention on the Law of the Sea, written in 1982 and entered into force in 1994, states that "the coastal [nation] exercises over the continental shelf sovereign rights for the purpose of exploring it and exploiting its natural resources." This provision is somewhat weakened by other provisions related to the concept of Exclusive Economic Zones, but the definition and extent of continental shelves remains an important concept in international law.

CHAPTER 4

THE MIRACULOUS MOLECULE

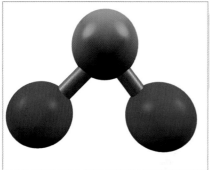

Left: Sculpted icebergs, the ocean reflecting sunset colors, and the clouds above are all made of water. Top: The attraction of one water molecule for another creates the surface tension that holds water droplets together. Bottom: Each water molecule has just three atoms. This illustration shows the typical form created by water's two hydrogen atoms (blue) and central oxygen (green) atom.

Water's chemical structure is quite simple: two hydrogen atoms bound to a single oxygen atom. Yet this simple structure gives water the remarkable physical and chemical properties that make it unlike any other substance on Earth.

Because of the way the atoms fit together, one end of a water molecule has a slight negative charge, and the other a slight positive charge. This makes each water molecule a tiny magnet that can attract a wide variety of other molecules, the property that lies behind water's ability to dissolve so many substances. Each water molecule sticks to other water molecules as well, creating a hydrogen bond. This bond creates a sort of skin on the surface of the water, a contracting force known as surface tension, without which there would be no waves. A slight change in the angle of this hydrogen bond when water freezes makes ice less dense than water so that it floats. Even something as basic as the fact that your clothes get wet if you jump in the water is a product of the molecular structure of water.

Floating and Sinking

Water exists in three forms: liquid, solid, and gas. Pure water's solid state, ice, occurs at temperatures of 32°F (0°C) and below. The gaseous state, vapor, requires temperatures of 212°F (100°C) and above. What makes water different from most other substances is that its solid form is less dense than its liquid form. That is, ice floats. The fact that it does has tremendous implications for climate and life on Earth.

To understand why floating ice is remarkable, let us start by looking at density. Density is a measure of how much a certain volume of a substance weighs. One cubic centimeter of water weighs 1 gram (0.035 ounce), while the same volume of air weighs only 0.0012 gram (0.000042 ounce). Water is much denser than air. But the density of water is not always the same.

TEMPERATURE

The exact density of water depends on its temperature. As water heats up, the molecules move more quickly, bouncing off one another and making the fluid take up more space. Therefore, warmer water is less dense than cooler water. Indeed, as early as the seventeenth century, English physicist Robert

Above: In coastal wetlands, fresh and salt water come together, creating layers of water with different densities. Top left: Water vapor, which is invisible, can be "seen" as the gap between the spout of a kettle and the millions of tiny droplets of liquid water that form the white cloud we think of as steam.

Hooke worried that as water temperature increased toward the equator, its density would decrease enough that ships might float noticeably lower. He warned captains sailing south from the Arctic to be careful not to overload their ships. While no shipping disasters are known to have occurred due to density changes in seawater, the fact that cold water sinks is essential to the global current patterns that drive Earth's climate.

But something strange happens as water gets close to the freezing point. As the molecules slow down and get closer

to one another, each water molecule can form hydrogen bonds with four other water molecules. The angle of these bonds widens slightly (a mere 4 degrees) when water changes from liquid to solid, and the water molecules form a crystal lattice. A space that would hold 27 molecules of liquid water holds only 24 of ice, meaning that a given volume of water will expand by 9 percent or so as it freezes.

SALINITY AND PRESSURE

Because salts are denser than water, salt water is denser than freshwater. While freshwater at 39°F (4°C) has, by definition, a density of 1 gram (0.035 ounces) per cubic centimeter (0.06 cubic inch), seawater at the same temperature has a density of 1.0278 grams (0.036 ounces) per cubic centimeter. This might seem like a small difference, but it is enough to make freshwater float on top of salt water. You can see this phenomenon in action where a muddy river meets the sea. The brown river water will float above the clearer seawater.

In the 1800s, some scientists believed that pressure had such a strong influence on the density of water that "at a certain depth the sea is specifically heavier than any body which we are acquainted with, consequently a cast-iron shell could not penetrate it." It is now clear, however, that pressure has only a small effect on water density, since water is nearly incompressible.

When seawater first freezes, it forms delicate "frazil" crystals like these. Frazil ice can give the ocean an oily appearance.

FREEZING WATER FASTER

There is a common belief that hot water freezes more quickly than cold. In closed containers, this is impossible. All water must cool to the freezing point to form ice, and hot water has much farther to go than cold water to get to that point. In 1969, however, Canadian G.S. Kell demonstrated that with water in uncovered buckets, the common belief holds true.

Kell set out two open buckets of water, one hot and one cold, on a freezing night. Since there were no lids, the water was able to evaporate, and hot water evaporates faster than cold water. Therefore, the bucket of hot water lost about 16 percent of its original volume to evaporation, leaving much less water to freeze. And of course as the water evaporated, it took heat with it, which also sped up the cooling process.

Does this mean you should use hot water when you want ice cubes quickly? Probably not. You would end up with smaller ice cubes, a warmer freezer, and a higher electric bill.

Starting with hot water will result in smaller ice cubes.

The Global Thermostat

Water has a higher heat capacity than almost any other substance on Earth. That is, it can take in or give up a tremendous amount of heat without changing its own temperature. Anyone who has waited impatiently for a cup of hot tea to cool down or for a pot of water to boil has an instinctive understanding of what this means. On a global scale, the heat capacity of water means that the ocean can release heat to colder air or take up heat from warmer air without any drastic change of its own temperature. This keeps the temperature difference between day and night from being too extreme, moderates the summer heat and winter cold, and accounts for the relatively mild climates of coastal areas compared to those inland.

The difference between temperature and heat is worth clarifying. Temperature is a measure of how fast, on average, the molecules in a particular substance are moving. A higher temperature means the molecules are moving faster, while a lower temperature means slower movement. Heat, in contrast, depends not just on how fast the molecules are moving, but also on how many of them there are. Thus, a cup of tea and a bathtub full of very hot water may have the same temperature, but the tub will have more heat, since it has a larger volume of water.

Left: This thermogram (yellow and orange are hot, purple and blue are cold) shows heat being transferred from a kettle into a teacup. Right: Bubbles form in boiling water when the water goes from liquid to a gaseous state. The air in the bubbles is pure water vapor. Top left: Coastal areas experience milder climates than inland areas due to water's tendency to warm up and cool down much more slowly than land or air.

HOLDING HEAT

If you put water in the freezer, its temperature will drop. When it is cold enough, the water will begin to solidify, to turn to ice. At this point, the temperature of the water will stop dropping, even as more heat is removed. For every hydrogen bond that forms as water turns to ice, heat is created and must therefore be removed. In fact, to turn just 1 gram of liquid water at 32°F (0°C) into ice takes 80 calories of heat. This is the same amount of heat it takes to increase the temperature of water from 32° to 176°F (0° to 80°C). Removing 80 calories from boiling water would create a noticeable drop in temperature, while removing 80 calories from liquid water at freezing would not change the temperature at all. Freezing cold water can get colder only once it is all frozen. At that point, removing more heat once again can make the ice colder.

RELEASING HEAT

A similar phenomenon happens when water changes from liquid to vapor. Add heat to water in its liquid stage, and the temperature goes up until you get to the boiling point, 212°F (100°C). At this point, the temperature of the water will stay the same until all the water has evaporated. To turn 1 gram of liquid water at 212°F (100°C) into vapor takes a whopping 540 calories. This so-called latent heat of evaporation is high because the hydrogen bonds holding water together are quite strong. For water to turn to vapor, all those bonds must be broken, which takes a good deal of energy. That is why sweating cools you down: As the sweat (water) evaporates, or turns to vapor, it takes heat with it.

The removal and release of heat as water evaporates and condenses is another way that water moves heat between the atmosphere and the ocean. As the Sun heats the surface of the ocean, water evaporates, taking heat with it. When the water vapor condenses again, forming clouds or fog, it releases heat. In addition to regulating temperatures on Earth, heat energy is part of what powers storms, winds, and currents.

Altocumulus clouds like these form between 6,500 and 20,000 feet (1,981–6,096 m) above the ground. They are usually made of water droplets but may contain ice if temperatures are cold enough.

The Universal Solvent

Water is capable of dissolving more substances—gases, liquids, and solids—than any other common liquid. Why is this? Once again, the answer has to do with the structure of the water molecule. Because one end of the water molecule has a negative charge and the other a positive charge, it can attract both negatively and positively charged molecules, or ions, and separate molecules that are held together by the attraction of negative and positive charges. Table salt, for instance, is made up of a negatively charged chloride ion and a positively charged sodium ion. Water molecules pull the sodium and chloride apart, clumping around each ion.

WHAT IS IN IT?

Even "pure" seawater is not just water. About 3.5 percent of it is composed of the dissolved salts that make it, well, salty. Although sodium chloride, or table salt, is the most well-known component, there are many others. For every 100 gallons (379 liters) of seawater that evaporate, in addition to almost 24 pounds (11 kg) of sodium chloride, there remain about 2.2 pounds (1 kg) of sulfate, 1 pound (0.45 kg) of magnesium, and more than 5 ounces (142 g) each of calcium and potassium.

Almost every element found in Earth's crust and atmosphere is also found in seawater, albeit in very small quantities.

The ocean's saltiness, or salinity, varies from place to place. When freshwater input is large, as happens near big rivers or melting sea ice, salinity is lower. Salinity along parts of the North American coast, for instance, is

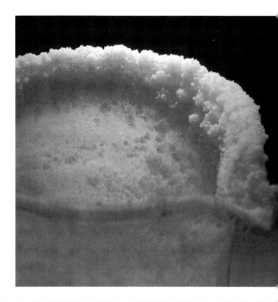

Top: Photograph of sodium chloride crystals. A crystal's shape reflects the arrangement of the ions that make it up. Bottom: Researchers being lowered onto sea ice, where they will measure light absorption, salinity, nutrients, and biomass in both the ice and the freshwater melt ponds on top of it. Top left: A high evaporation rate coupled with relatively low freshwater input has given the landlocked Dead Sea the saltiest water on Earth. Water in the upper hundred feet or so is more than 30 percent salt, ten times saltier than the ocean.

commonly around 3.3 percent. In warmer areas with a great deal of evaporation, such as in the tropics, salinity may be as high as 3.7 percent.

Salinity affects many of water's properties. More salt means a lower heat capacity; that is, the water will heat up more quickly. Add salt, and the freezing point lowers: Seawater can remain liquid down to almost 28.4°F (-2°C). Since salt molecules attract water, seawater evaporates more slowly than freshwater.

Salinity also affects the ability of water to transmit electricity. Pure water does not conduct electricity (but pure water is very hard to get, so do not try this at home!). The saltier water is, the better it conducts electricity. This fact gives oceanographers an easy way to measure salinity: By measuring how well water conducts electricity, they can tell how salty it is. Conductivity meters are the most common way of measuring salinity in oceanographic research.

WHY SO SALTY?

The volume of the ocean is huge, and so is the volume of salt it contains. Some people estimate that the oceans contain enough salt to form a layer 500 feet thick around the entire planet. Why have the oceans accumulated so much salt, while lakes and rivers remain relatively fresh? Rivers carry water and dissolved salts to the ocean; underwater volcanoes and hydrothermal vents add salt to the ocean as well. Even the air can bring salt to the oceans, as winds carry particles from land out over the seas. While water evaporates from the surface of the ocean and returns to the land as rain, the salts stay behind. This same phenomenon accounts for the high salinity of lakes that no longer have an outlet, such as the Dead Sea or the Great Salt Lake. Oceans only lose salt when certain organisms use it to make shells or other hard body parts.

WHY ARE SOME ICEBERGS GREEN?

Most icebergs have a bluish tinge, but their apparent color may change as the quality of incoming light changes. Icebergs containing gravel or dust may appear brown. Some icebergs, however, are a deep green. Just why the ice attains this color remains a mystery.

Most scientists agree that green icebergs begin as a layer of frozen seawater on the bottom of an ice shelf. That is where the agreement ends. Some researchers suggest that the color comes from high levels of certain metals trapped in the ice, but chemical analyses have refuted this claim. Others suggest that the green results from red light interacting with blue icebergs; at least in some cases, however, the green is clearly part of the iceberg itself. Still others suggest that the real culprit is dissolved organic material that is released by phytoplankton growing next to the ice. While this is often the case, it does not seem to be universally true. And so the mystery continues . . .

Light and Dark

The deep blue sea is not always blue. It ranges from the crystal blue waters around a tropical reef to the mysterious dark of the deep ocean and the cloudy green of algae-rich coastal water. Behind these many hues lies the way light is reflected, absorbed, or scattered by the water and the organisms and particles suspended in it.

What we call light is one part of the electromagnetic spectrum. The color of light is determined by its wavelength. The color of an object or a liquid is determined by the wavelengths that it reflects or scatters back to our eyes.

Ultraviolet (UV), X-rays, and gamma rays all have wavelengths too short for our eyes to pick up, while infrared, microwaves, and radio waves have wavelengths that are too long for us to see.

THE OCEAN BLUE

The blue of the open ocean tells us that it absorbs longer wavelengths of light most readily, leaving only the shorter blue wavelengths to be scattered and reflected back to us. In fact, if a person cut her hand just a mere

Left: Even in the clear waters of the Caribbean, underwater photographers still need lights to get good images. Right: In a so-called red tide, vast numbers of microscopic organisms called dinoflagellates give the water its red color. Dinoflagellates can produce toxins that are taken up by animals that feed on them, making it dangerous to eat clams, mussels, and certain other animals that have experienced a red tide. Top left: The water around this tropical island is turquoise blue because it contains relatively small amounts of algae, sediments, or other materials that would influence the way the water absorbs light.

30 feet (10 m) below the surface, her blood would not look red at all: No red photons make it more than a few meters below the surface. By 350 feet (106 m), orange and yellow light have been completely absorbed, and then by 500 feet (152 m), almost all the green light is gone as well.

Both the quantity and the color of light change dramatically with depth. In the clear open ocean, 10 percent of incoming light is absorbed within the first 246 feet (75 m), and 99 percent within the first 500 feet (152 m). Below 3,300 feet (1,006 m), there is too little sunlight for even the most sensitive of fish eyes to detect. Eyes are still useful, though, because of the widespread ability of living things to produce their own light, a phenomenon known as bioluminescence.

Anything that is added to water changes which wavelengths are absorbed, which are scattered, and which are reflected. The silt from rivers or stirred-up sediments scatters more light and makes the water brown; red tides are caused by immense numbers of microscopic organisms that reflect back the red wavelengths. It is the relative absence of any suspended matter or even microscopic life that makes tropical water such an alluring blue.

Water clarity is often measured using the simplest of tools: Secchi disks, which are white or black-and-white disks lowered into the water on a rope or cable. The depth at which the disk disappears from view

A researcher uses a Secchi disk to measure water clarity. These disks were invented in the late 1800s by astrophysicist Pietro Angelo Secchi, a scientific adviser to the pope.

is called the Secchi depth and is used to compare water clarity around the world. Other, more precise measurements of clarity are achieved by measuring how much light is absorbed by a small sample of water.

IT LOOKS DIFFERENT TO ME

Oceanographers Tammy Frank and Edith Widder were surprised to find that a species of deep-sea shrimp they were studying seemed to respond very specifically to UV radiation. Previous research had shown that shorter wavelengths of light are typically absorbed rapidly in seawater, and people had long assumed that UV radiation was completely absorbed within the first few feet. These shrimp

lived hundreds of meters below the surface, so why would they have the ability to perceive UV? Curious, Frank and Widder rigged up a submersible to measure light as it descended. It turns out that in the clear water around the Bahamas, some UV makes it down to 1,700 or even 2,000 feet (518 or even 609 m)! It is not a lot of UV, but it is enough for the shrimp to notice.

Even though we cannot see UV, it is an essential part of the visual spectrum for many animals. With UV, many of the tiny animals that live in the water shine brightly; without it, they are dark specks that blend into the background. In the absence of UV, the larval stages of some species of fish are unable to feed well enough to survive.

Sound

For those of us who live in a world of light, vision often seems more important than sound. But for whales, it is sound that matters. While light gets rapidly absorbed by water, sound does not. Sound travels five times faster and sixty times farther in seawater than in dry, room-temperature air, and the low-pitched calls of some great whales can travel great distances. Whales and other animals use sound to determine distance, direction, and the properties of the world around them. By using images produced by sound, we have made what was once invisible visible. While deep ocean basins are beyond the reach of sunlight, they are well within the reach of sound.

Above: A researcher in the Mediterranean lowers a sonar probe into the water to search for archaeological remains on the seafloor. Top left: Orcas, like most toothed whales, use sound as well as vision to "see" their environment. Because of this, high-powered military sonar can be damaging to them. Possible effects include disorientation, which can lead to stranding.

ECHOES IN THE SEA

Sound travels in waves, and like all waves, its speed and direction can be altered by the medium through which it travels. When sound hits a hard surface like a submarine or a fish, it reflects off of it, or echoes. This simple principle is behind what is now an indispensable tool on seagoing ships: sonar (sound navigation and ranging). If you know how fast sound travels (4,921 feet/1500 meters per second in seawater), you can tell how far away something is by measuring the time between when you emit a noise and when the echo reaches you.

Sound does not reflect off all surfaces equally, though. By analyzing the sounds that return, oceanographers can classify the seafloor. Using low-frequency sound waves that penetrate the seafloor as well as the water above it, oceanographers can "see" not only the seafloor but also what lies beneath it. This technique can show the contours of hard rock underlying sediments.

When sonar became common in World War II, sailors noticed an odd "false bottom" or "deep scattering layer" on their sonar screens. This layer seemed to move up at night and down during the day, and submarines would use it to hide themselves. This layer, it turns out, is created by huge numbers of fish and other organisms that migrate daily through the water column.

SO NEAR AND YET SOFAR

Sound waves refract, or change direction, as they pass through layers with different properties. They generally bend toward areas where their velocity is lower. Increasing temperature and pressure increase the velocity of sound, a phenomenon that leads to some interesting ocean acoustics, since temperature tends to decrease with depth, while pressure increases.

If the surface water is relatively well mixed, there is usually a velocity maximum of around 350 feet (100 m) below the surface. Above this layer, sound waves bend upward, while below it they bend downward. This phenomenon creates an acoustic shadow zone: Beyond a certain distance from the source, no sound waves will reach the area just under the zone of maximal velocity. Submarines can take advantage of this shadow zone to hide from ships on the surface.

There is also a distinct zone of minimum sound velocity, usually between 2,000 and 4,000 feet (609 and 1,219 m) depth in low and midlatitudes. Closer to the surface, the speed of sound is strongly influenced by temperature; sound waves bend downward. Farther down, the effect of pressure comes into play, and sound waves bend upward. In the thermocline, the area where temperature changes rapidly with depth, sound waves bend down as temperature decreases, then bounce back up as pressure increases, only to bounce back down again. This zone is called the sofar (sound fixing and ranging) channel, and sounds emitted here travel particularly long distances. Explosions in Australia can be heard as far away as Bermuda.

Submarines can hide in the deep scattering layer of the ocean.

USING SOUND TO MEASURE CLIMATE CHANGE

All else being equal, warmer water transmits sound faster than cooler water does. Scientists are taking advantage of the sofar channel and this simple phenomenon to calculate shifts in ocean temperature. By measuring changes in how long it takes sound to travel from special emitting stations near Hawaii and San Francisco to stations as far away as New Zealand, scientists can tell how much the temperature has changed. Using this method, scientists can detect temperature differences as small as 0.0108°F (0.006°C). Although researchers were concerned that the sounds might disturb marine mammals that live near the emitting stations, whales and seals do not make any effort to avoid them.

Scientists prepare to install a sound source in the Arctic Ocean.

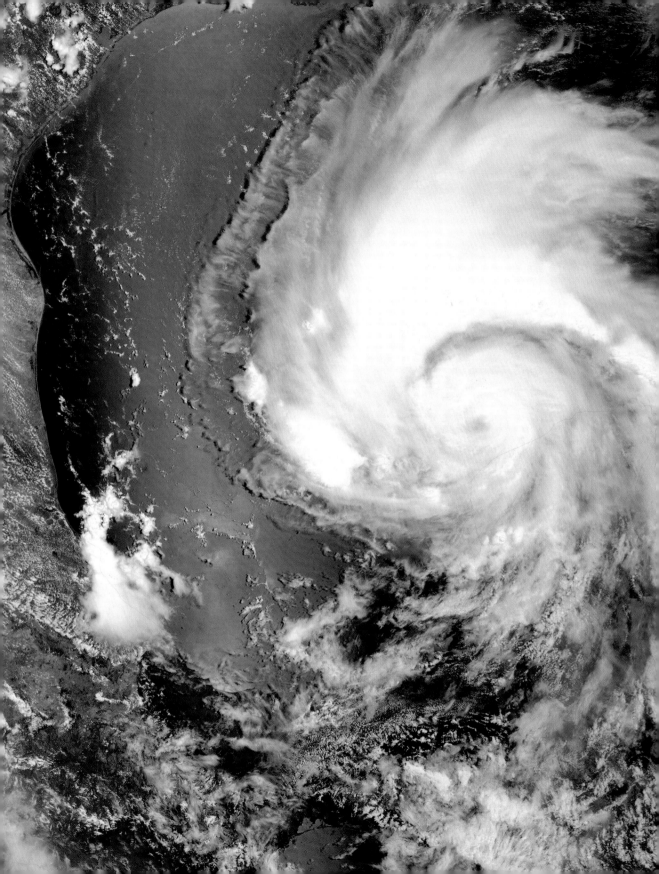

CHAPTER 5

THE AIR ABOVE

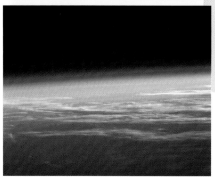

Left: This satellite image was taken on July 18, 2005, as Hurricane Emily traveled over the Gulf of Mexico toward the U.S. shoreline. Top: Interactions between air, land, and sea are behind the conditions that have sculpted this windswept tree on the Canadian coast. Bottom: New satellite technology allows researchers to see Earth's atmosphere from the side rather than from above. The stratosphere, mesosphere, and thermosphere form a blue halo over the clouds.

The oceans and the atmosphere are intimately connected. Both readily absorb and transfer heat. Both are influenced by differences in density and in fact have several layers of different densities within them. Forces generated by our spinning Earth alter the paths of air and water, and waves travel readily through both. Although air and water behave differently in many ways, the principles underlying major patterns of air and water movement are surprisingly similar.

The interplay among air, land, and water is responsible for creating powerful hurricanes and gentle sea breezes. Variations in the amount of solar radiation that regions receive—which areas heat up or cool down faster—create the trade winds, the monsoons, and the doldrums. Understanding the principles that control atmospheric movement has allowed us to generate short-term weather forecasts. In an effort to understand the forces that drive longer-term weather and climate patterns, scientists are creating and testing innumerable models, which may help to predict and prepare for shifts brought about by the current global climate change.

Layers of Air

The atmosphere can be defined as a layer of mixed gases extending from Earth's surface out into space. On average it is 78 percent nitrogen, 21 percent oxygen, and 1 percent other gases. The concentration of gases, and thus the atmosphere itself, becomes less and less dense the farther away from Earth it gets. The atmosphere is divided into four layers based on how temperature changes with altitude.

FOUR LAYERS

The lowest layer, the troposphere, is warmed by Earth's heat, so temperature decreases rapidly at higher altitudes. Anyone going up a tall mountain can notice this. The thickness of the troposphere varies from around 10 miles (16 km) in the tropics to 6 miles (10 km) over the poles. The troposphere contains 80 to 90 percent of the atmosphere's total mass and is where weather happens. It is capped by the tropopause, a region where the air is a stable but chilly -76°F (-60°C).

The next layer, the stratosphere, is where the UV-shielding ozone layer is found. Because ozone absorbs energy from the Sun so effectively, and more energy is available at higher altitudes, the strato-

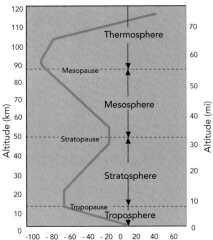

Above: Dinoflagellates are marine organisms that produce dimethylsulfide, a sulfur-containing gas that affects climate over the ocean. Top left: Every year millions of house-size comets break up and release clouds of water vapor in Earth's upper atmosphere. Like all comets, these are made primarily of water, but their lack of dust and iron prevents them from having the bright tail of larger comets, like Hale-Bopp, shown here.

sphere gets warmer rather than colder with increasing altitude. The stratopause, another region of stable temperature, marks the top of the stratosphere, about 30 miles (50 km) above Earth.

Next is the mesosphere, where temperature once again decreases with altitude. Because the mesosphere has little ozone to trap heat, temperatures may get as low as -173°F (-114°C). The mesosphere is capped by the mesopause at 55 miles (85 km) above the Earth's crust.

The outermost layer of Earth's atmosphere is the thermosphere. Temperature in this layer again increases with altitude, but here molecular

A schematic of the atmosphere, showing how the relationship between temperature and elevation changes in each layer. The thermosphere extends much farther and gets much hotter than shown but has no sharp upper edge because the concentration of gas molecules decreases so gradually.

oxygen rather than ozone heats things up. Although temperature in the thermosphere can be higher than 2,700°F (1,500°C), the density of molecules is so low that it would feel quite cold. The different layers within the thermosphere enable radio waves to reflect back to Earth, which made long-distance radio communication possible before the advent of satellite technology.

Intense solar radiation in the thermosphere splits gas molecules into charged particles called ions. The region of the thermosphere filled with these particles is called the ionosphere, and ions are what create the dazzling high-latitude light displays known as auroras.

WARMING THINGS UP

Some of the gases in the troposphere are called greenhouse gases because they keep Earth warmer than it would be without them. They let most incoming solar radiation pass through and warm the Earth, but then trap the heat that radiates back out from the planet. The most common, although not the most powerful, greenhouse gas is carbon dioxide. Other greenhouse gases include water vapor and ozone. Concentrations of many greenhouse gases have increased dramatically as a result of human activity, which is probably one reason why Earth has gotten measurably warmer in the last century. Without any greenhouse effect, Earth would be too cold for human life, but too much greenhouse effect may create a world unlike any we have known before.

COOLING THINGS DOWN

While greenhouse gases keep things hot, other substances in the atmosphere cool things down. Sulfur compounds and other tiny particles in the troposphere reflect solar energy back to space before it has a chance to warm the planet. These cooling substances come from volcanoes, from pollution, and, amazingly, from microscopic marine organisms called dinoflagellates. Ships have been shown to produce a significant amount of sulfur pollution: Patterns of tropospheric sulfur over the ocean closely follow patterns of shipping activity.

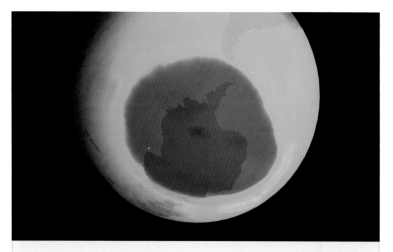

The dark blue region over Antarctica is the ozone hole, where ozone concentrations have become abnormally low.

LOSING OUR SUNSCREEN

Since the 1980s, scientists have tracked the seasonal appearance of holes in Earth's protective ozone layer. The biggest and most predictable is over Antarctica, but holes also appear in the Arctic and temperate latitudes. The international 1987 Montreal Protocol successfully reduced the concentration of many ozone-depleting chemicals. Almost 200 countries have agreed to abide by the Montreal Protocol, an international agreement phasing out the use of ozone-depleting chemicals. While the protocol has been very successful in some ways, a number of ozone-depleting chemicals are still in use.

Earth's protective ozone layer is further depleted by an increase in greenhouse gases. The destruction of ozone happens most effectively when the stratosphere is very cold. By trapping more heat in the troposphere, greenhouse gases keep the stratosphere cooler. The loss of ozone itself also cools the stratosphere, since absorption of solar energy by ozone is what warms the stratosphere.

What Lies Behind the Ice Ages

Earth's climate cycles from warm to cold, wet to dry, over periods ranging from years to decades to millennia. Some of these cycles are well understood, while others are not. Serbian scientist Milutin Milankovitch proposed one of the most elegant theories of cyclical change in the early 1900s, linking changes in Earth's orbit around the Sun with ice ages. The three elements of Earth's orbit on which Milankovitch focused were how tilted Earth's axis is relative to the Sun (obliquity), how circular Earth's orbit is (eccentricity), and which part of Earth is closest to the Sun at which time of year (precession). These are called the Milankovitch cycles.

THE CYCLES

The most familiar effect of the Earth's tilt is the occurrence of seasons. The hemisphere that is tilted toward the Sun experiences summer, while the other experiences winter. As Earth moves around the Sun, the hemisphere facing the Sun alternates, and so do the seasons. With no tilt, Earth would have no seasons, and the greater the

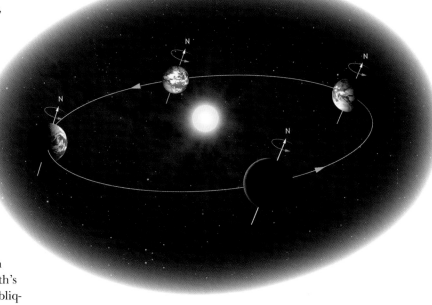

Earth's tilt, the greater the difference between seasons. Every 41,000 years, the tilt of Earth's axis cycles between 22 and 29 degrees away from vertical. Right now, Earth is in the middle of a cycle, with an axial tilt of 23.5 degrees, meaning fewer seasonal differences.

Earth's orbit around the Sun is oval, not circular, so Earth is closer to the Sun during

Above center: Changes in which hemisphere faces the Sun cause the annual climate cycles we know as seasons (illustrated). Milankovitch cycles operate on cycles of thousands of years and affect the climate of the entire Earth. Above right: Serbian mathematician Milutin Milankovitch. Top left: Large boulders may be carried far from their source by glaciers, then left behind when climate changes and the glaciers melt.

some parts of the orbit than others. While this does not determine when summer and winter happen, it does influence how extreme the seasons are. Right now, Earth is closer to the Sun in January, which means that Northern Hemisphere winters are milder and Southern Hemisphere summers hotter. Eleven thousand years ago, when Earth was closest to the Sun in July, the situation was reversed, as it will be in another 11,000 years.

The longest Milankovitch cycle measures just how circular Earth's orbit is. The period of this cycle from most circular to least circular is 95,000, 125,000, and 400,000 years. Right now, Earth is five million kilometers closer to the Sun in January than it is in July, which translates into a 6 percent difference in how much solar radiation reaches Earth from season to season. When Earth's orbit is at its most oval (most eccentric), the difference can be as high as 30 percent.

MILANKOVITCH AND CLIMATE

Milankovitch theorized that all of these cycles in Earth's orbit combine to determine how much solar radiation reaches Earth, which in turn is the driving force

for cycles in ice ages. Because most land is in the Northern Hemisphere, Milankovitch also guessed that what happens in the Northern Hemisphere is most important in starting or ending ice ages.

Cooler northern summers would prevent ice from melting and lead to the formation of larger ice sheets. Larger ice sheets would reflect more light back to space, cooling Earth even further. In 1976, five decades after Milankovitch first put forth his theory, deep-sea cores confirmed that climate cycles for the past 450,000 years matched Milankovitch's predictions.

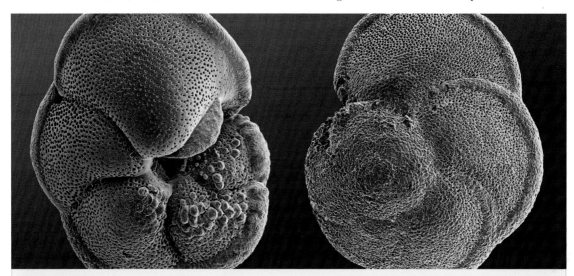

Many marine organisms, like these foraminifera, incorporate oxygen from the water into the calcium carbonate of their shells.

OXYGEN TELLS ALL

Oxygen naturally occurs in three forms, or isotopes, called oxygens 16, 17, and 18. The numbers indicate molecular weight—oxygen 18 is the heaviest, oxygen 16 the lightest. Many marine organisms incorporate oxygen from the water into the calcium carbonate they use to make their shells and teeth, but how much of each isotope they use depends on the water temperature. This means that scientists can use the ratio of oxygen 18 to oxygen 16 found in the hard parts of some marine organisms as a sort of thermometer. In fact, scientists have used this method to measure ocean temperatures from the present to more than 100 million years ago. By looking at the oxygen isotope ratios in fossils of organisms that lived on the seafloor and organisms that lived throughout the water column, researchers can even study how differences in surface and bottom water temperature change over time.

Regional Climate Seesaws

While some climate cycles occur on the scale of thousands or hundreds of thousands of years, others happen over years or decades. The most famous of these is the El Niño–Southern Oscillation, which can affect climate across much of the globe for 6- to 18-month periods.

Under normal conditions, there is a low-pressure area across Indonesia and the western Pacific along the equator. A region of high pressure sits over Easter Island and the eastern equatorial Pacific. This difference in pressure is what fuels the trade winds: Air moves from the high-pressure eastern Pacific to the low-pressure western Pacific. The wind and air-pressure patterns keep sea levels about 20 inches (50 cm) lower along Central and South America than in the western Pacific. In a process called upwelling, these patterns create conditions in which cold, nutrient-rich water rises up along the western coast of Central and South America. These nutrients feed blooms of photosynthetic organisms along the coast, which in turn support some of the most abundant fisheries in the world.

Every few years, these conditions are reversed. The trade winds switch direction, and warm water piles up along the

Above: This false-color image shows wind speed and direction over the ocean. Orange indicates higher wind speeds, and blue lower. The image was generated using a SeaWinds scatterometer that measures the scattering and reflection of microwave radar by the ocean's surface. Top left: A NOAA employee services one of over 70 buoys moored in the equatorial Pacific. These buoys provide real-time ocean temperature information that helps scientists predict El Niño events.

west coast of South and Central America. Average seawater temperature goes up several degrees, and sea level rises by 12 inches (30 cm) or so. Upwelling stops, and the productivity of these coastal ecosystems drops. Because this phenomenon often occurs around Christmas, it was dubbed El Niño, the Christ Child. The reversal of these normal atmospheric pressure conditions is the Southern Oscillation, so scientists often use the term

ENSO, for "El Niño–Southern Oscillation," to refer to these atmospheric changes and their biological effects.

FAR-REACHING EFFECTS

Because of the complex atmospheric and oceanic processes that determine weather around the globe, the effects of ENSO can be widespread. It leads to unusually warm winters in the northwestern United States and Canada. In the eastern United

States and usually dry parts of Peru and Ecuador, conditions become much wetter than normal, while typically wet regions along the western side of the Pacific become drier. During extreme ENSOs, the changes in currents and increase in ocean temperature along the eastern Pacific can bring tropical animals as far north as Alaska.

DECADAL CYCLES

Like ENSO, other climate cycles have received attention because of their effects on fisheries. Fisheries biologist Steven Hare's investigations during the 1990s linked fluctuations in salmon production in Alaska to regional climate. This discovery led to the description of the Pacific Decadal Oscillation, or PDO. The PDO switches between warm and cool phases every few years or decades. When the PDO is in a warm phase, fisheries off Alaska do well while coastal fisheries farther south suffer; in a cool phase, the situation is reversed.

Other major cycles include the Arctic–North Atlantic and Antarctic Oscillations, which many scientists feel control climate at high latitudes. While they may flip between warm and cool phases in a matter of weeks or months, they have both been primarily in their warm, or positive, phase for the past several decades. Just what controls ENSO, PDO, and the polar oscillations is unclear. Volcanic eruptions, ozone depletion, and greenhouse gas levels have all been suggested as possible contributors, but for now the mystery remains unsolved.

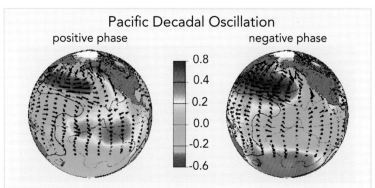

Pacific Decadal Oscillation
positive phase   negative phase

0.8
0.4
0.2
0.0
-0.2
-0.6

This diagram shows winter sea surface temperature anamolies, wind patterns (arrows), and sea level pressure (contour) for the warm (left) and cool (right) phases of the Pacific Decadal Oscillation.

FORECASTING THE FUTURE?

The global climate is changing. Most places are getting warmer, although some are getting cooler. Patterns of rain and snowfall are changing. Scientists and others are wondering what this means for Earth's future. Because recorded human history contains no similarly dramatic change in the global climate, answering this question can be tricky.

Some scientists are using the Pacific Decadal Oscillation to help. In the past century, there have been two warm and two cool phases. This alternation of phases lets scientists look at what happens to currents, upwelling patterns, and ecosystems along the west coast of North America during extended periods of warmer-than-usual weather. Although the PDO is smaller in scale than global climate change, studies of its effects can help scientists prepare for what global change may bring.

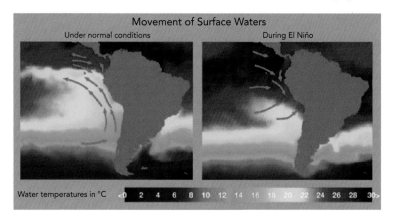

Movement of Surface Waters
Under normal conditions   During El Niño

Water temperatures in °C   <0 2 4 6 8 10 12 14 16 18 20 22 24 26 28 30>

Normally, air pressure in the eastern Pacific, which is higher than that in the west, fuels winds that drive coastal upwelling. This brings cold, nutrient-rich waters up along the coast of North and South America and supports productive fisheries. During El Niño years, this gradient wanes or reverses, inhibiting upwelling and the fisheries that depend on it.

The Coriolis Effect

In 1835, French physicist and engineer Gustave-Gaspard Coriolis published a paper explaining how the laws of motion applied to rotating bodies. Although he himself never connected his work with the circulation of the ocean or atmosphere, William Ferrel demonstrated that the Coriolis effect was responsible for the curvature of weather fronts at higher latitudes. It deflects movement to the right in the Northern Hemisphere and to the left in the Southern Hemisphere. Just how much deflection occurs depends on the speed and distance traveled, and the distance from the equator. Contrary to popular belief, the Coriolis effect does not generally determine which way water spins as it goes down the drain. It does, however, play a key role in many of the wind, weather, and current patterns around the globe.

DEFLECTION OF NORTH-SOUTH TRAVEL

Pick any city on the equator—Quito, Ecuador, for instance. Like everything else on Earth's surface, it makes one complete revolution around Earth's axis every day. The Earth's circumference at the equator is about 24,902 miles (40,075 km), so someone standing still in Quito is spinning around Earth's axis at 1,037 miles per hour (1,669 km/h). Someone on the Arctic Circle is not moving nearly as fast. The circumference here is about 10,975 miles (17,662 km). Since it still takes 24 hours to complete a revolution, someone on the Arctic Circle moves along at just 457 miles per hour (736 km/h), less than half the speed of someone in Quito.

What if someone in Quito shot a missile due north? The force with which it was fired would determine its northward speed, but its east-west speed is determined by how fast it was spinning around Earth's axis when it was fired. The missile would continue to zip east at more than 1,000 miles an

Above: Because this low-pressure system occurred in the Northern Hemisphere (Iceland is on the right), the Coriolis effect caused winds and clouds to spiral counterclockwise as they moved from areas of high pressure to areas of low pressure. Top left: One of NOAA's "Hurricane Hunters" in flight. Teams of scientists and engineers fly through hurricanes to collect meteorological information used to predict the hurricane paths.

hour (1,609 km/h) as it flew north. The eastward velocity of the Earth below it would get progressively slower, so the missile would be moving east faster and faster relative to the ground below.

From space, the missile would appear to travel in a straight line, but to an observer in Quito, its path would seem to curve to the east, or right. Similar visualizations illustrate why a ball would also appear to curve right if shot from the Arctic to the equator, and why it would appear to curve left in the Southern Hemisphere.

THE DEFLECTION FOR EAST-WEST TRAVEL

What if someone north or south of the equator fired a missile due east? The relative speed of the surface of the Earth at different latitudes would not matter, yet the ball would still be deflected to the right in the Northern Hemisphere and to the left in the Southern Hemisphere. Why?

An object traveling in a circle experiences two distinct forces: centripetal force pushing inward, and centrifugal force pushing outward. If the moving object changes speeds, then the balance of these two forces shifts. Speeding up pulls the object farther away from the center of rotation, while slowing down causes it to "fall" toward the center.

Objects on Earth's surface are really spinning around in big circles, with Earth's axis as their center of rotation. A missile fired due east moves faster relative to

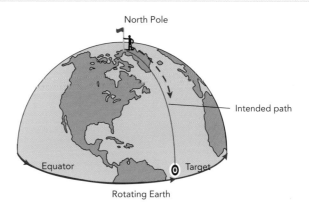

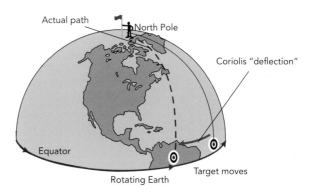

Although the underlying physics are different for objects traveling north-south and east-west, the Coriolis "deflection" is always to the right in the North and to the left in the South. This diagram shows how Earth's rotation makes an object no longer fixed to the Earth appear to deviate from its original path.

Earth's axis than the ground underneath. Centrifugal force will pull the missile outward. Since gravity keeps it from zipping off into space, the only way the missile can get farther away from Earth's axis is by moving toward the equator, that is, to the right in the north and the left in the south. A missile fired due west would move more slowly than the Earth below, so the missile would move closer to Earth's axis. In this case, that means moving toward the pole, or, once again, to the right in the north and the left in the south.

Carried by the QuikSCAT satellite, SeaWinds can monitor winds over the ocean in all kinds of weather.

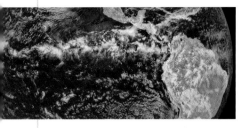

Whichever Way the Wind Blows

Wind is just air moving from one place to another. Air flows from areas where pressure is higher to areas where pressure is lower. These differences in air pressure usually come from differences in temperature. Warm air expands and rises, creating a low-pressure area that air from surrounding regions moves in to fill. What creates the heat differences that underlie global wind patterns? To a large extent, these differences are caused by how much solar radiation is absorbed at different latitudes. Near the equator, sunlight shines almost straight down. This shortens its path through the atmosphere, so less sunlight is absorbed before reaching Earth. Also, the atmosphere itself is thinner near the equator than at the poles. The relatively vertical path of light near the equator also means that the energy from a given amount of light is spread over a much smaller area than it would be at higher latitudes. Finally, because high latitudes have more land and ice, they reflect more incoming radiation than lower latitudes, which have more open ocean. All of this explains the obvious: The equator is much warmer than the poles!

MAJOR WIND PATTERNS

The temperature differences across latitudes cause air to rise at the equator, carrying heat and moisture with it. At the tropopause, the air stream's upward motion stops, and it begins moving north or south. As this air flows toward the poles, the Coriolis effect deflects it to the east, creating the eastward jet streams around 30 degrees north or south of the equator. The air then cools and compresses as it moves away from the equator, until at around 30 degrees north or south of the equator it sinks back to Earth.

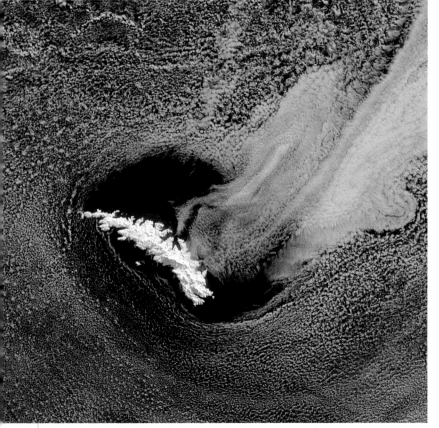

Above: South Georgia Island is framed by a break in the clouds. Cloud formations like those shown here are created by different wind patterns. Convection cells are indicated by the lacy pattern at the upper left of this image. Top left: The stream of clouds over the equator shows where the trade winds come together.

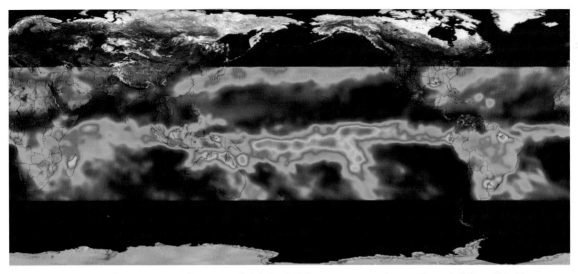

This composite image shows a one-month average of global rainfall measurements. Areas of low rainfall are light blue. Regions with heavy rains are orange and red.

This sinking air creates a band of high pressure, the "horse latitudes," where winds are often weak and unpredictable. The name may come from the fate of horses on board sailing ships at these latitudes. Poor winds could becalm boats for weeks, and if drinking water ran low, horses were thrown overboard. Another low-pressure region around the equator also has generally low and unpredictable winds; this region is called the doldrums.

Once air at 30 degrees North or South reaches the Earth's surface, it heads toward the equator or the poles. Air flowing toward the equator creates the trade winds, which, because of the Coriolis effect, blow in a northwest or southwest direction. Air flowing toward the poles creates winds that blow from west to east. Because winds are named for the direction from which they come, these winds are known as the westerlies.

At 60 degrees North or South, air once again rises, levels out at the tropopause, and flows north or south. There is a second eastward-blowing jet stream, and air sinks either around the poles or at 30 degrees North or South. Air flowing outward from the poles creates the polar easterlies.

All of these large, circular flows of air are called convection cells. The convection cells that run from the equator to 30 degrees North or South are known as Hadley cells, those in the mid-latitudes are the Ferrel cells, and the high-latitude ones are the polar cells.

RAIN

The air that sinks at 30 degrees North or South is dry, which, along with high pressure, tends to keep skies clear. This is why many of the Earth's major deserts occur at this latitude. At the equator and 60 degrees North or South, the rising air warms and moistens the region, creating tropical and temperate rain forests. These precipitation patterns affect the oceans as well. In the tropics and at 60 degrees North or South, the heavy precipitation lowers surface salinity noticeably, while the desert conditions at 30 degrees North or South create saltier seas.

A wind-powered vessel of the nineteenth century. International trade and early ocean exploration were directly affected by wind patterns.

Sea and Land Breezes

Water heats up and cools down much more slowly than land. When the Sun rises, the land heats up faster than the ocean does, and the air above the land will heat up sooner than the air over the ocean. The warm air rises, creating lower pressure over the land. Cooler, denser ocean air moves into this low-pressure area, creating an onshore, or sea, breeze. The breeze picks up as the day progresses, peaking in midafternoon, and wanes as land and sea temperatures equalize again. At night, the land cools more quickly than the sea, and the pattern reverses, creating an offshore, or land, breeze.

These breezes can be strong, carrying the smell of land 19 miles (30 km) or more offshore. The pattern of land and sea breezes can also influence the habits of fishermen who rely on wind to carry them out to fishing grounds in the morning and back home at night. These breezes are even responsible for the famous Bay Area fog in coastal California, where local currents often create a band of cold water close to shore. As the sea breeze pulls warm air over this water in the morning, the air cools, and water vapor condenses, creating fog. Fog may also form in the evening, when

Top: Twin typhoons spin over the Pacific Ocean near the Philippines. Bottom: The rain shadow effect is powerful enough to see from space. The leeward, southwest sides of the Hawaiian Islands are noticeably browner than the windward, northeast sides. Top left: San Francisco's famous fog results from the cooling of warm air as it moves over cold coastal waters.

warm air from the land travels back out over the cold water.

BIG RAINS

Monsoons are seasonal winds that are, much like sea and land breezes, driven in part by differences in the rate at which land and water heat up and cool down. Although monsoonlike wind patterns occur around the globe, the term is most frequently used to refer to wind patterns in the Indian Ocean and southern Asia. The word "monsoon" has its roots in the Arabic word *mausin*, which means season.

While sea and land breezes result from daily changes in temperature, monsoons result from seasonal changes. In summer, large landmasses warm quickly relative to the oceans, and air is pulled across the oceans as the warm inland air rises. The ocean air brings moisture, generating heavy seasonal rains. While often causing disastrous floods, monsoon rains are essential for successful agriculture throughout Asia. In winter, the situation reverses, bringing dry winter monsoon winds. Indian monsoons blow from the southwest in summer and from the northeast in winter.

RAIN SHADOWS

For most people, the words "Hawaiian Islands" conjure up images of a lush, wet, tropical paradise. In fact, only the sides of the islands that are exposed to the wind, called the windward sides, have these charac-

Dimethyl sulfide (DMS) is one of many compounds that influence the formation and appearance of clouds. Fine ash and aerosols produced by powerful volcanic eruptions can lead to startlingly red sunsets.

WHAT IS THAT SMELL?

People often comment on the smell of the sea, but just what *is* that smell? One major component is dimethyl sulfide (DMS), which plays a role not only in the smell of the sea but in climate as well.

Certain types of phytoplankton, microscopic photosynthetic organisms, produce a chemical precursor of DMS. When damaged, these phytoplankton release their contents into the surrounding water, where bacteria and other organisms convert the precursor into DMS and other compounds. Some DMS diffuses or gets sprayed into the air, where it forms little particles which may play a role in seeding clouds. DMS also changes the reflectivity of clouds: Clouds with a lot of DMS or other sulfur particles reflect more solar radiation. This process is usually self-controlling. More clouds mean that less light and heat reach the ocean surface, which reduces the growth rates of the photosynthetic organisms that control levels of DMS in the first place.

teristics. As the wind comes off the ocean, it hits the mountainous island interior. This increases air pressure, which decreases air's ability to hold moisture. By the time the winds reach the sheltered, or leeward, sides of the islands, most of the moisture is gone. This phenomenon occurs wherever there are large mountain ranges. The Amazon basin, on the windward side of the Andes, is one of the wettest places on Earth, while the leeward side, along the west coast of South America, has places where it rains only once every few years.

Storms

The devastation caused by Hurricane Katrina along the Gulf Coast of the United States in the summer of 2005 was stunning. Winds greater than 135 miles per hour (217 km/h) destroyed trees and houses and created a deadly storm surge, lifting sea level almost 30 feet (9 m) above normal. Every year, hurricanes and other storms wreak havoc in coastal regions around the world. A storm surge in the Bay of Bengal in 1970 killed an estimated 300,000 people.

CATEGORIES OF STORMS

In the words of the National Weather Service, a cyclone is "a large-scale circulation of winds around a central region of low atmospheric pressure, coun-terclockwise in the Northern Hemisphere, clockwise in the Southern Hemisphere." Cyclones are most common in the tropics but do occur outside the tropics, even in the Arctic.

Tropical cyclones with sustained wind speeds less than 38 miles per hour (62 km/h) are known as tropical depressions. They have no eye and often are not spiral in shape. If sustained wind speeds are between 39 and 73 miles per hour (62–117 km/h), the event becomes a tropical storm. Storms with sustained wind speeds greater than 74 miles per hour (119 km/h) get different names depending on where they happen. In the Atlantic basin and the eastern North Pacific, they are called hurricanes. In the western North Pacific and much of the South Pacific, they are called typhoons. Over the Indian Ocean and parts of the South Atlantic, they may be called severe cyclonic storms, severe tropical cyclones, or simply cyclones.

Hurricanes are further categorized according to wind speed using the Saffir-Simpson scale. The least powerful have wind speeds between 74

Above: A tropical depression moves across the United States. Cooler-than-usual water and surrounding thunderstorms robbed this former hurricane of much of its power. Top left: A 1958 diagram showing a cross section of a hurricane.

and 95 miles per hour (119–153 km/h) and storm surges of 4 to 5 feet (1–1.5 m). The most powerful, category 5 storms, have winds greater than 155 miles per hour (249 km/h) and storm surges higher than 18 feet (5.5 m). Flooding may cut off low-lying escape routes several hours before the worst of the storm arrives.

EYES AND EYE WALLS

In hurricanes, air is pulled along the Earth's surface toward the central low-pressure region. As the air rushes inward, the

This photograph, taken by a hurricane hunter pilot, shows Hurricane Katrina's eye wall the day before the storm hit the U.S. Gulf Coast. At this point, Katrina was a category 5 hurricane.

Coriolis effect deflects winds to the right in the Northern Hemisphere and to the left in the Southern Hemisphere, creating the classic spiral shape. More powerful cyclones also have well-developed eyes, which are areas of relative calm up to 30 miles (50 km) across in the middle of the storm. Just outside the eye is the eye wall, where both wind and rain are most powerful, whipping around the eye at furious speeds.

STORMS AND GLOBAL WARMING

Tropical cyclones get their energy from the ocean. The warmer the temperature of water is, the more energy it has to feed the cyclone. Because of this, several computer climate models predict that global warming will create bigger hurricanes. However, evidence from the past is mixed. Peter Webster of the Georgia Institute of

Technology and his colleagues have found a global increase in the number of category 4 and 5 storms over the past 35 years, but this time scale is too short to pin the blame securely on global warming. Roger Pielke from the University of Colorado has observed that where and how towns and cities are built may matter more in determining just how much destruction will be caused by future hurricanes.

Had Hurricane Frances occurred in the South Pacific rather than the Atlantic, it would have been called a typhoon.

WHAT'S IN A NAME?

Cyclone, typhoon, and hurricane are different words for the same type of storm. The word "cyclone" comes from the Greek word *kyklos,* meaning "circle," which may be related to the Egyptian word *cykline,* meaning "to spin."

The source of the word "typhoon" is hotly debated. It may come from the Greek mythological character Typhon, father of dangerous winds, or from *tufun,* a word that occurs in Arabic, Persian, and Hindi. Given that it was used to refer specifically to storms in the South China Sea, many scholars favor a Chinese origin, the Formosan term *t'ai fung,* meaning "eminent wind."

On the other side of the world, the word "hurricane" came to the Europeans from the Taino people of Haiti. It probably arose first in South America and shares its origin with the Mayan god of wind and storm, Hurakan. Although hurricanes are associated with destruction, Hurakan was a creator, drying out the ocean with his breath to create the land.

CHAPTER 6

WATER ON THE MOVE

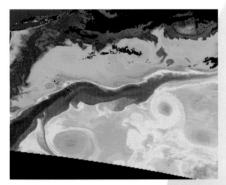

Left: A view from space of the Gulf Stream as it separates from the North American coast at Cape Hatteras, North Carolina. Top: The Gulf Stream travels up the East Coast of the United States, heading across the Atlantic Ocean from just south of the Chesapeake Bay to the British Isles. This false-color image shows the difference in temperature between the stream and the surrounding waters. The coldest waters are shown as purple, with blue, green, yellow, and red representing progressively warmer water. Bottom: Traigh Beach in the West Highlands of Scotland is located along the path of the Gulf Stream.

The study of ocean currents matters for a wealth of reasons, including their effects on commerce, climate, and national security. Matthew Fontaine Maury pioneered the use of ships' logs to create maps of global wind and current patterns. Maury's 1855 book, *The Physical Geography of the Sea*, captures the poetry as well as the science of the sea:

> There is a river in the ocean. In the severest droughts it never fails, and in the mightiest floods it never overflows. Its banks and its bottom are of cold water, while its current is of warm. The Gulf of Mexico is its fountain, and its mouth is in the Arctic Seas. It is the Gulf Stream. There is in the world no other such majestic flow of waters. Its current is more rapid than the Mississippi or the Amazon.
>
> Its waters, as far out from the Gulf as the Carolina coasts, are of an indigo blue. They are so distinctly marked, that their line of junction with the common seawater may be traced by the eye. Often one half of the vessel may be perceived floating in the Gulf Stream water, while the other half is in common water of the sea, so sharp is the line, and such the want of affinity between those waters, and the reluctance, on the part of those of the Gulf Stream to mingle with the common water of the sea.

The Gulf Stream is just one of many currents flowing through the ocean. A complex interplay of wind, gravity, and other factors shapes these currents that, in turn, affect commerce, climate, and life in the seas in innumerable ways.

Invisible Boundaries

The ocean is divided into many regions, each with its own set of physical, chemical, and biological properties. These regions may be separated by physical barriers such as an isthmus or an underwater ridge, or by differences in the water itself, such as density or currents. Oceanographers call a body of water with physical and chemical characteristics distinct from the water around it a water mass. These water masses may be transient—for example, a lens of freshwater on the surface that gets mixed in during the next storm; they may also be long-lived and stable. Persistent water masses are often given names, such as the Antarctic Bottom Water or the Arctic Intermediate Water.

DENSITY

Normally, less dense fluids float on top of more dense fluids. This phenomenon is called density stratification, and it is visible in all kinds of fluids, including the Earth's mantle, a container of salad dressing, and the oceans.

At the surface of the ocean is a zone called the mixed layer. Here wind and waves stir up the water and make temperature and salinity fairly constant throughout the zone. The mixed layer is usually about 500 feet (150 m) thick but may be as much as 3,200 feet (1,000 m) thick—or it may be virtually absent. It is thicker in turbulent conditions and thinner when waters are calm.

Below the mixed layer is the pycnocline, where water density changes rapidly with depth. Under the pycnocline is intermediate water, reaching down to 5,000 feet (1,500 m), and below that is the deep water, which goes down to about 12,800 feet (4,000 m). The lowest layer is the bottom water, the densest of all.

STABILITY

If the density difference between layers of water is great, the water column is more stable and less likely to experience vertical mixing. This is because the greater the density difference, the more energy it takes to push surface water down or pull deeper water up. Think of the difference between trying to hold a ball filled with air underneath the water and trying to hold a ball filled with mud underneath the water. It is much harder to get the less dense air-filled ball to stay down.

The size of the density difference is affected by changes in the density of the surface water. Increased evaporation makes surface water saltier, increasing its density and decreasing the density difference between the mixed layer and the water below. Increased rainfall decreases

Above: The Southern Ocean, which surrounds Antarctica, holds the cold, dense water known as Antarctic Bottom Water. The cooling of the ocean and formation of sea ice during winter increases the water's density, so it sinks into the deep sea. Top left: The Atlantic coast of the Western Sahara photographed from space. Increased evaporation in areas like this makes the surface water much saltier than normal.

An aerial view of the frozen Southern Ocean showing an area of open water where the frozen surface has been pulled apart by currents and wind. These areas, called leads, can meander for miles.

salinity, making the surface less dense and increasing the density difference.

There are often seasonal trends in surface density. At higher latitudes, surface water will cool considerably in winter. If it cools enough, it becomes denser than the water below it and starts to sink. Water can also change temperature and density if it is carried from a warmer region to a cooler region or vice versa.

CURRENTS

Planktonic organisms by definition are weak swimmers at best, and where they go depends primarily on where the currents take them. Because mixing between water layers is rare, plankton may be essentially trapped above or below the pycnocline. Strong currents may also create barriers. For instance, there is a major shift

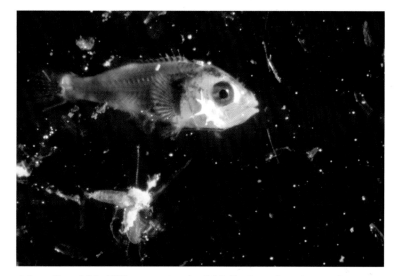

A juvenile rockfish. Different species of rockfish live in various ocean regions, from shallow coastal waters to deeper waters on the edge of the continental shelf.

in intertidal species north and south of Point Conception in southern California. At Point Conception the shoreline bends sharply to the east, and the southward-flowing California Current moves offshore. Many marine animals release their larvae into the water column for dispersal to new locations, but larvae coming from the north will be swept away from suitable habitat. Larvae floating up from the south would not be able to swim against the current.

Upwelling and Downwelling

Many of the world's most productive fisheries are in areas characterized by coastal upwelling. In these areas, the regular influx of nutrient-rich water from below the surface fuels the growth of algae and other photosynthetic organisms, just the way adding fertilizer to fields increases plant growth. This increase in primary productivity supports an abundance of animal life. Around Antarctica, coastal upwelling is what makes the ocean so rich in krill—the shrimplike animals essential to the diets of penguins, whales, and many other animals. Because of the interaction of wind and water, upwelling zones are most common along the western edges of continents.

WINDS AND CURRENTS

If surface water is pushed away from shore by the wind, deep water will rise up to replace it. Although it might seem as though offshore winds would be behind the major upwelling systems, these systems occur in areas where wind blows parallel to the shore. This is because the interaction of the Coriolis effect and the drag of wind on water or of one layer of water on another creates a net flow of surface water perpendicular to the direction of the wind.

Along the west coast of South America, the predominant winds blow north along the coast. Surface water is deflected to the left in the Southern Hemisphere, so there the winds generate a strong offshore current. Along the west coast of North America, the winds are from the north, water is deflected to the right, and once again an offshore current is formed.

Upwelling is variable in space, time, and intensity. In some areas, upwelled water stays close to the coast. In other areas, it can head offshore in extended plumes or filaments. Along the northern California coast in North America, some plumes extend more than 120 miles (200 km) out into the ocean.

WHY SO RICH?

As plants, algae, and other photosynthetic organisms grow in the surface water, they use up the nutrients in that water. When organisms die, they usually sink to the bottom and take their nutrients with them. Without any input of new nutrients, the growth of the primary producers rapidly becomes limited. Tropical and open oceans are often nutrient poor, which is what makes the water such a clear blue: There are

Above: A phytoplankton bloom off the coast of Argentina is visible in this image as brilliant swirls of green and blue. This region of the Atlantic Ocean, where cool, nutrient-rich water from the Antarctic mixes with warm, salty water flowing south from Brazil, is particularly productive. Top left: Humpback whales feeding on krill.

fewer primary producers to make it appear cloudy.

Deeper in the water, the absence of sunlight reduces primary productivity, which means that fewer nutrients are used up. This, combined with nutrients drifting down from above, means that deeper water is generally much richer in nutrients than are surface waters. Upwelling brings these nutrients back to the sunlit zone, where plants, algae, and other organisms can once again make use of them.

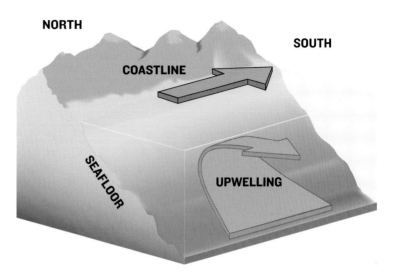

UNEXPECTED CONSEQUENCES

Many marine organisms float freely with the currents, unable to exert much control over their location. Even such stay-at-home animals as barnacles and clams have young that are cast adrift for days, weeks, or months before they can settle down to the sedentary life of an adult. The strong offshore currents associated with upwelling can move these young so far away from the coast that they never make it back. The ability of certain larvae to survive into adulthood depends on the temporary relaxation of upwelling conditions in their habitats.

Upwelling supports some unexpected industries. Along the west coast of Peru, large populations of fish draw great numbers of birds to the area. The birds then deposit nitrogen- and phosphate-rich droppings known as guano on the rocky shoreline, where people collect it and use it to make fertilizer.

Above: Barnacles on the shores of San Juan Island, Washington. Top: The process of upwelling along the western coast of North America. Strong winds blowing north to south (red arrow) generate offshore flow of surface water, and deep, nutrient-rich water rises to replace it.

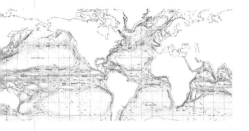

Rivers on the Surface of the Sea

The surface of the ocean is crisscrossed with wind-driven surface currents. The trade winds create the west-flowing North and South Equatorial Currents, while the midlatitude westerlies push water east, forming the North Pacific and North Atlantic Drifts in the Northern Hemisphere and the West Wind Drift in the Southern Hemisphere. Closer to the poles, the polar easterlies create the East Wind Drift in the Southern Hemisphere and the Oyashio Current in the Northern Hemisphere.

AROUND THE OCEANS

East-west currents interact with continents and gravity to create north-south currents. For instance, the North Equatorial Current piles water up in the western Pacific, creating a pressure gradient that pushes water north along Japan, forming

the Kuroshio Current. On the eastern side of the Pacific, the California Current carries water south, while the Alaska Current heads north. The California Current flows into the North Equatorial Current, completing a circle of currents around the North Pacific. Major north-south currents in the South Pacific are the Peru or Humboldt Current along western South America, and the East Australian Current.

The North and South Atlantic likewise are each bounded by a circle of currents. In the North Atlantic, the east-west currents are matched by the northeast-flowing Gulf

Stream and south-flowing Canary Current. In the South Atlantic, the Benguela Current flows north along western Africa, while the Brazil Current carries water south. Some of the South Equatorial Current is deflected north as well, crossing the equator and eventually joining the Gulf Stream.

SPEED

Surface currents, averaging between 0.3 and 1.5 feet (0.1 and 0.5 m) per second, move at about 1 percent the speed of the winds 30 feet (10 m) above the surface. Current speed changes with the width and depth of the current.

Above left: An equatorial wave line in the Pacific Ocean. This line is an equatorial boundary between warm surface water and cool, recently upwelled water. Above right: The Florida Current speeds up as it squeezes between Florida and Cuba. Top left: Major ocean surface currents are visible as areas of warmer water, as here are the Agulhas Current (south of Africa), the Kuroshio Current (off Japan), and the Gulf Stream (off North America).

Where the Florida Current squeezes between Florida and Cuba, for instance, it speeds up to 5 feet (1.5 m) per second.

In a phenomenon known as western intensification, currents along the western edge of oceans are narrower, faster, deeper, and more sharply defined than those along the eastern edges. This is the result of a number of factors. In addition to the equatorial easterlies piling water up along the western edge of ocean basins at low latitudes, planetary waves known as Rossby waves are constantly moving from east to west across the oceans, influencing the flow of currents. Western intensification is much weaker in the Southern Hemisphere than in the Northern Hemisphere because of the different arrangement of continents.

EDDIES AND WINDROWS

The interaction of a fast-moving current with the slower water around it creates swirls of water that break off from the main flow of the current, called eddies. These eddies take with them the characteristics of the current from which they come. Surface eddies near the southeastern United States carry the chemical signature of water from Gibraltar, 2,500 miles (4,000 km) away. Eddies vary from a few to several hundred miles across, can last from days to weeks, and occur at all depths.

Sometimes, long lines of bubbles or other floating materials stretch out on the ocean's surface parallel to

Franklin and Folger's map of the Gulf Stream. Franklin was intrigued by the idea of a stream existing within an ocean. On a 1775 voyage from England to the colonies, he took temperature measurements of the ocean as often as four times a day.

SLOW ROUTE FROM ENGLAND

In the years preceding the Revolutionary War, Benjamin Franklin was postmaster general for the American colonies. In those days, mail traveled by boat, and mail boats going from Falmouth, England, to New York took weeks longer than boats going from London to Rhode Island. Franklin wanted to know why. He turned to Timothy Folger, a cousin and a Nantucket ship captain, for advice. Folger said that the problem was the Gulf Stream, a powerful wind-driven current flowing northeast from America to Europe. The Falmouth–New York boats were traveling directly against the current, while the London–Rhode Island boats crossed over it. The Gulf Stream mattered not only for efficiency of sailing but for finding whales as well. These mammals tended to congregate along the edges of the Gulf Stream where prey was plentiful. Because the Gulf Stream was so critical for New England boat captains in those days, Folger could sketch its path and give directions for avoiding it from memory.

the wind. These result from shallow horizontal spirals of current known as Langmuir cells. Side-by-side cells rotate in opposite directions, creating convergence and divergence zones between the cells. Where the downward-flowing sides of cells meet, a convergence zone is formed. Water from each side of this zone carries buoyant material with it, but when the water flows downward, the buoyant material is trapped at the surface, creating those parallel lines on the surface.

Gyres and Spirals

When Norwegian explorer Fridtjof Nansen froze his boat in the Arctic pack ice in the 1890s to track ocean currents, he noticed that the direction of drift was usually 20 to 40 degrees to the right of the wind direction. He surmised that this was the result of the Coriolis effect: Since Nansen was in the Northern Hemisphere, as the wind pushed the ice in one direction, the Coriolis effect deflected it to the right. He suggested that below the ice, water would move even more to the right of the wind direction. As the ice, moving at an angle to the wind, pulled on the water below it, that water would move not in the direction of the ice, but to the right of that direction. This water would pull on the water below it, which would also be deflected to the right. The result would be a gradual spiraling of current direction with depth. After hearing a talk by Nansen on this subject, Swedish oceanographer Walfrid Ekman was so intrigued that he promptly set to work on the physics and mathematics behind it. In 1902 he published a theoretical account of the phenomenon, which now bears his name, the Ekman spiral.

HILLS IN THE WATER

One fascinating result of the Ekman spiral is that in the upper 500 feet (150 m) or so of the water column, the net movement of water is 90 degrees to the right or left of the wind direction, depending on which hemisphere it is measured in. That is, water in the surface layer of the ocean moves at right angles to the wind. This phenomenon is called Ekman transport, and because of it, there are noticeable "hills" in each of the major oceans.

The hills form at the center of gyres, regions outlined by roughly circular currents. For instance, the north-flowing Gulf Stream, the east-flowing North Atlantic Current,

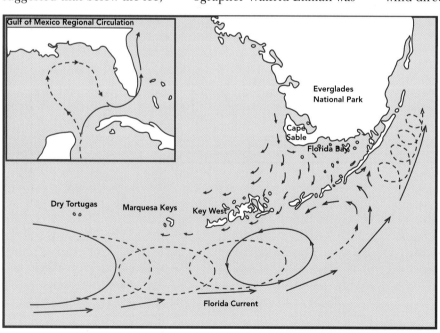

Above: This map shows patterns of currents through the Florida Keys. Numerous gyres exist between the path of the Florida Current and the shoreline. Top left: Fridtjof Nansen's ship, the Fram, *photographed in January 1912.*

A large spiral eddy in the Sea of Japan. Eddies are common in this area where the Tsushima Current, a small branch of the Kuroshio Current, travels between Japan and Korea and onward to the Pacific Ocean.

the south-flowing Canary Current, and the west-flowing North Equatorial Current create the North Atlantic gyre. The east-west currents are driven by the dominant winds, generated by the differential heating of Earth at different latitudes, and the Coriolis effect, a result of Earth's spin. The north-south winds are driven by a pressure gradient created by the east-west winds. For instance, the North Equatorial Current piles water up along the Atlantic coast of North America, which then pushes water north in the Gulf Stream.

Because the currents of the North Atlantic gyre circle clockwise, Ekman transport dictates that there will be a net movement of surface water into the center of the gyre. Indeed, water at the middle of the gyre may be 3 feet (1 m) higher than water at the edges. Pressure from this piled-up water pushes water out from the center of the "hill," but the Coriolis effect deflects it to the right, that is, around the hill. The two forces roughly balance out, and the result is a surface current flowing around the base of the "hill."

THE BIG CIRCLES

There are five major gyres: the North Pacific and South Pacific gyres, the North Atlantic and South Atlantic gyres, and the Indian Ocean gyre. These gyres circle consistently in the same direction: those in the Northern Hemisphere circle clockwise and those in the Southern Hemisphere circle counter-clockwise. Smaller gyres driven by seasonal wind patterns may reverse direction periodically. Gyres in the Arabian Sea, for instance, may change direction when the monsoon winds reverse direction.

The Great Ocean Conveyor Belt

Some have called the great ocean conveyor belt the largest river in the world. Moving at a stately 4 inches (10 cm) per second around the globe, it has a total volume 100 times that of the Amazon River. Carrying heat from equatorial regions north and south along the surface, it warms western Africa and Europe. Water that cools and sinks near the poles travels the ocean's depths until it rises, more than 1,000 years later, in the Pacific or Indian Ocean.

THAT SINKING FEELING

When surface water becomes denser than the water below, it sinks. It may get denser by cooling, as when seasons change. It can also become denser by getting saltier, as happens in areas of high evaporation. The formation of sea ice also increases salinity since water, but not salt, is incorporated into the ice. Because the conveyor belt is driven by these changes in temperature and salinity, it is also called the thermohaline circulation.

There are two key areas where dense water forms and sinks. One is the Norwegian Sea, where warm water flowing north from the equator meets the Arctic sea ice. The surface water gets colder and saltier and sinks 3,000 to 12,000 feet (1,000–4,000 m) to form the North Atlantic Deep Water. It then flows south across the bottom, around South Africa, and into the Indian and Pacific oceans. The densest water in the ocean, the Antarctic Bottom Water, is formed in the Weddell Sea off the coast of Antarctica.

What goes down must come up. Although the upwelling zones are not as clear as the downwelling zones, there seem to be areas of diffuse upwelling in the Pacific and Indian oceans. Scientists map the upwelling of deep water using its distinctive chemical signature.

DEEP-SEA MASS EXTINCTION

Around 55 million years ago, many deep-sea species became extinct in a short span of time. Research has revealed changes in types of organisms and in their shell chemistry, which indicates a rapid shift to warmer, oxygen-poor water, perhaps due

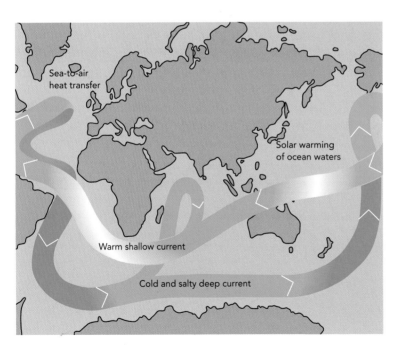

Sea-to-air heat transfer

Solar warming of ocean waters

Warm shallow current

Cold and salty deep current

Above: A simplified illustration of the thermohaline circulation, also known as the great ocean conveyor belt. Top left: A false-color image showing concentrations of chlorophyll in the Indian Ocean on February 1, 2002. Chlorophyll is found in phytoplankton, which are most highly concentrated in areas where upwelling occurs.

to a shutdown or reversal of the thermohaline circulation. What could have caused such a massive change in ocean circulation? One likely culprit is runaway global warming caused by the release of massive amounts of methane, a powerful greenhouse gas, from frozen deposits on the ocean floor. Other evidence points to plentiful rain and little ice, so high-latitude surface waters were most likely fresher than normal, slowing or preventing the formation of dense water in those areas. Also, the large but shallow equatorial Tethys Sea may have experienced enough evaporation that its surface water became salty enough to sink to the ocean floor. Had dense water been produced at the equator rather than at high latitudes, the conveyor belt would have run in reverse.

FUTURE SHUTDOWN?

The release of water from two major glacial lakes into the North Atlantic 8,200 years ago cut the conveyor belt's speed in half for close to a century. The conveyor belt seems to have shut down almost completely at least twice between 20,000 and 10,000 years ago. In both cases, it seems that melting icebergs created a thick lens of freshwater on top of the ocean that halted normal downwelling processes. When the conveyor belt shuts down, temperatures around the North Atlantic drop precipitously.

In our time, there is evidence that both sea ice and glaciers are melting at a rapid pace due to the current warm-

ing trend. This has scientists worried that a conveyor belt shutdown could happen again soon. Indeed, in 2005 a British group reported that the speed of the conveyor belt in the North Atlantic had slowed by 30 percent over the past 50 years.

Above: Ice floes in the northern Bering Sea. Researchers use melting sea ice as evidence of changes to the Earth's climate. Top: An eight-week-old Weddell seal surfacing from under pack ice. These seals live in the densest water in the ocean, which is formed in the Weddell Sea.

Modeling the Oceans

Understanding something as complex as ocean circulation can be a daunting task. As oceanographers expand their investigation of currents and circulation, computer models have become an indispensable addition to their toolbox.

Computer models not only provide a way to visualize and present complex three-dimensional data, but also can suggest possible trends or patterns that scientists might not notice, and highlight areas where more data would be particularly useful in resolving unknowns. They are used for activities as diverse as predicting the path a hurricane will follow, understanding the interaction of sea surface temperature and climate change, and determining the best place to locate a sewage treatment plant. They also allow oceanographers to answer "what-if" questions that they hope will never be tested in the real world. For instance: What if there is a major oil spill in the Gulf of Alaska? And what happens if the spill occurs in winter rather than summer, in calm rather than windy conditions? While the answers to these questions that technology offers may not be perfect, they are better than having no information at all.

REALITY CHECK

When scientists create models, they check them against actual data. To get a sense for how well a model predicts the interaction of a warming climate and ocean currents, for instance, modelers might "predict" what happened in the past. If the model correctly "predicts" the past, that is a good sign. If not, scientists must go back to the drawing board. Figuring out why a model was wrong means figuring out what we do not understand about how the oceans work. Models make oceanographers clearly express and test their hypotheses about mechanisms on a grand scale.

HURRICANES

The decision to evacuate or not in the face of a potential hurricane is tricky. If an area is evacuated unnecessarily, valuable time, money, and other resources are wasted. If an area is not evacuated when it should be, hundreds of people may be killed or injured. Predicting just how strong a hurricane will be and where it will go is very important.

Scientists from England and the United States have joined forces to help improve just such predictions. They are studying a number of variables—wind, wave-wave interactions, wave-bottom interactions, the loss of energy through whitecapping

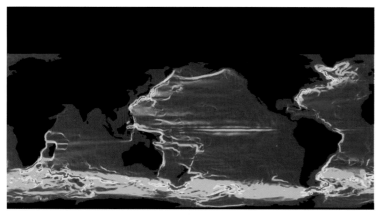

Above: A computer model of ocean currents derived from a simulation of ocean dynamics over four years. The colors represent water velocity from fastest (blue) to slowest (red). Top left: A researcher with the Scripps Institution of Oceanography prepares a water motion sensor, which examines surf zone water motion at La Jolla, California.

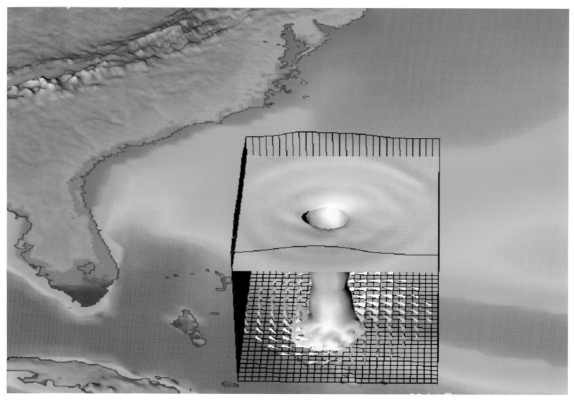

A computer model of Hurricane Floyd embedded within a global forecast model. Such models are used to investigate the impact of global climate on storm tracks and hurricanes.

on waves, and more—to figure out what makes hurricanes tick. By altering each variable in a computer model and comparing the results with the behavior of past hurricanes, these scientists hope to create a model that accurately tells people when to run.

VIRTUAL ESTUARY

In 1997, a group of educators and scientists got together to create a virtual version of Puget Sound, the second-largest estuary in the United States. They thought that it might be easier for students to understand the circulation of Puget Sound if they could manipulate a three-dimensional image of it themselves. Using a six-direction game controller, students and researchers alike can dive into the water, fly above it, and move in whatever direction takes their fancy, all the while looking at what is happening with the currents. The project was a success, and the Virtual Puget Sound is now part of a larger group of Puget Sound models generated for research and education. These models are being used to understand how the Puget Sound ecosystem works, to address and prevent environmental problems, and to inspire and educate people about the wonders of physical oceanography.

A 2002 satellite image shows characteristics of the lands surrounding Puget Sound. Dense urban areas are shown in black.

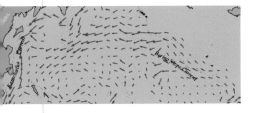

Studying Currents

Given the importance of ocean currents for climate, marine ecosystems, and travel, it is not surprising that humans have developed a plethora of approaches for studying currents over the centuries. Most of these methods fall into one of two camps: either they follow a particular parcel of water as it moves, or they measure how quickly water passes by one particular point.

The former technique usually involves tracking the movement of a floating object. Matthew Fontaine Maury had sailors throw bottles overboard throughout their journeys. Each bottle contained a note saying where it had been thrown overboard and asking the finder to let Maury know where the bottle had washed ashore. Drift cards, pieces of wood printed with information on when and where they were released and how to report a finding, are another low-tech way of tracking currents. More expensive floats can be tracked with radar or radio-tracking devices, or are designed to float at specified depths, sending acoustic signals to the surface.

Measuring speed at a single location is usually accomplished with a current meter suspended from a float on the surface, or anchored to the bottom. Each current meter has a rotor, which measures current speed, and some sort of vane to measure current direction.

GOING WITH THE FLOW

In 1969, six researchers boarded a submarine named the *Ben Franklin* to float with the Gulf Stream. They began their journey at a depth of 500 feet (150 m) off West Palm Beach, Florida, and were tracked from the surface by two boats throughout their trip. In addition to tracking the flow of the current, they measured temperature, salinity, and density almost continuously. The voyage

Above: Scientists deploy an Acoustic Doppler Current Profiler in the Ross Sea off the coast of Antarctica. Top left: Current-tracking devices were used to make this map of current speed and direction (red arrows) in the Bering Sea.

was full of the unexpected. At one point, the *Ben Franklin* was carried 50 miles (80 km) away from the main track of the Gulf Stream by an eddy and had to be towed back. From this the researchers learned that eddies are not restricted to the surface waters, since the submarine was hundreds of feet down. They also discovered numerous uncharted hills, which would suddenly deflect the current—and the sub with it—up, down, and sideways. Although the team saw fewer animals than they had expected, they recorded whale songs and photographed a diversity of bioluminescent creatures, some new to science. The voyage of the *Ben Franklin* ended after 31 days, when the sub surfaced a few hundred miles south of Halifax, Nova Scotia, 1,650 miles (2,640 km) from where it had started.

THE SPEED OF SOUND

As anyone who has heard an ambulance or a train zooming by knows, motion can distort sound. If the sound source is moving toward the listener, the pitch becomes higher. If it is moving away from the listener, the pitch becomes lower. This is known as the Doppler effect, and oceanographers take advantage of it to measure the speed of currents throughout the water column. Using an Acoustic Doppler Current Profiler, scientists bounce sound waves off particles in the water and use the relative distortion of the original sound wave to tell how fast the particles are moving.

A shipment of bath toys was lost at sea in 1992. Oceanographers followed the movements of the rubber critters from Sitka, Alaska, to the Aleutian Islands.

RUBBER DUCKIES FOR SCIENCE

On January 10, 1992, a ship traveling through rough seas lost 12 cargo containers, one of which held 28,800 floating bath toys. Brightly colored ducks, frogs, beavers, and turtles were set adrift in the middle of the Pacific Ocean. Oceanographers promptly recognized the scientific potential of the accident and asked beachcombers around the Pacific to tell them when and where these toys washed up. After seven months, the first toys made landfall on beaches near Sitka, Alaska, 2,200 miles (3,540 km) from where they were lost. Other toys floated north and west along the Alaskan coast and across the Bering Sea, washing ashore in the Aleutian Islands and Kamchatka, Japan. Some toy animals stayed at sea even longer: They floated completely around the North Pacific gyre, ending up back in Sitka after traveling 6,800 miles (10,900 km). Oceanographer Jim Ingraham has created an accurate computer simulation of regional currents using data from spills like this as well as satellite-tracked buoys.

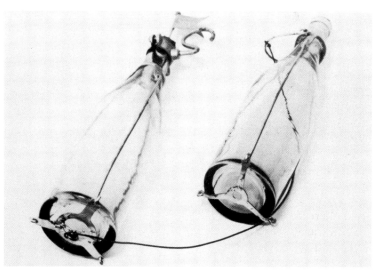

Alfred Hautreaux devised these bottle floats in 1893 to study currents in the Gulf of Gascogne.

Milestones in Ocean Science

4000 BCE
Polynesians engage in what may have been the first transoceanic voyages, exploring and colonizing islands across the Pacific.

3200 BCE
Egyptians and Mesopotamians use wind- and oar-powered reed boats for trade and travel along the Nile River, the eastern Mediterranean Sea, the Red Sea, and Arabian Sea.

2500 BCE
Map showing mountains and a river is created on a clay tablet in Babylonia.

1500–500 BCE
The Phoenicians, often considered the greatest of the early navigators, dominate trade in the Mediterranean. High-quality maps were key to their success.

1478 BCE
Queen Hatshepsut of Egypt sends out an organized expedition to explore the Red Sea and Somali coasts.

500 BCE
Phoenician navigator Hanno reputedly circumnavigates Africa.

450 BCE
Greek traveler and scholar Herodotus describes the regular tides of the Persian Gulf, uses the term "Atlantic" for the first time to describe the western seas, and makes a map of "the world," reflecting Greek belief that the oceans were a river that surrounded the land around the Mediterranean Sea.

360 BCE
In two dialogues written around this time, Plato tells the story of the lost civilization of Atlantis, a supposedly advanced civilization that disappeared beneath the waves over 9,000 years before his time.

325 BCE
Pytheas sails from Greece to northwestern Europe, calculating latitude by the North Star. He proposes that tides are controlled by the Moon.

240 BCE
Eratosthenes, a scientist at the great Library of Alexandria, calculates the circumference of the Earth with reasonable accuracy based on differences in the length of shadows at different latitudes.

150 CE
Ptolemy produces his famous map of the world, which served as a standard for over 1,000 years. Some of his mistakes remained in maps as late as the eighteenth century.

500
Polynesians are the first to settle in the Hawaiian Islands.

986
Erik the Red leads a group of Vikings from Iceland to a new land they call Vinland the Good, now known as Greenland.

Viking ship

1405–33
Chinese treasure ships traverse the waters between China and Africa. Many key maritime inventions—including the rudder, the compass, and the presence of multiple watertight compartments in the hold—trace their origin back to the Chinese.

1492
Columbus crosses the Atlantic and mistakenly believes the Caribbean Islands he reaches are part of India; his voyage signals the beginning of regular European travels to the Americas.

Columbus landing at San Salvador

1498
Vasco da Gama sails around Africa and on to India, opening a trade route between Portugal and India.

1520
Magellan begins his attempt at circumnavigating the globe. Although he dies en route, one of the ships from his party completes the journey.

1580
Sir Francis Drake leads the second circumnavigation of the globe, setting the stage for years of European fighting over who owns which territories in the New World.

1643

Evangelista Torricelli of Italy invents the barometer, an instrument used to measure atmospheric pressure. Barometers are used to track weather patterns and predict oncoming storms.

1685

Edmond Halley publishes a manuscript that describes patterns of trade winds and monsoons, links solar heating with atmospheric movement, and establishes the link between barometric pressure and elevation.

1687

Isaac Newton publishes *The Mathematical Principles of Natural Philosophy*, in which he lays out his laws of gravitation and uses them to create a general explanation of how tides work.

Sir Isaac Newton

1728–61

John Harrison of England develops several versions of a pendulum-free clock capable of keeping accurate time on a moving boat. This allows sailors to determine longitude as well as latitude, making global travel much easier.

1751

Henri Ellis measures deep-sea temperatures in the tropics and discovers a layer of cold water below the warm surface layer.

1768–79

Captain James Cook makes four celebrated voyages around the Pacific, charting winds, currents, depth, temperature, and coral reefs, bringing the existence of New Zealand, Australia, and the Sandwich and Hawaiian Islands to the attention of Europeans, and demonstrating the usefulness of Harrison's chronometer for global sea travel.

1775

Pierre-Simon Laplace publishes the dynamic theory of tides, which adds real world characteristics to Newton's ideal tide theory.

1769

Benjamin Franklin and Timothy Folger publish a map of the Gulf Stream. Although sailors had been aware of the Gulf Stream for years, the map allowed less knowledgeable captains sailing from Europe to the Americas to avoid it. Franklin published two later versions of the map in 1778 and 1786.

1831–36

HMS *Beagle*, under the command of Captain Robert Fitzroy and carrying Charles Darwin, travels extensively around South and Central America.

1843

Edward Forbes presents his "azoic theory," claiming that no marine life exists below 300 fathoms (549 m). Despite evidence to the contrary, the theory held sway for years. While Forbes was wrong on the deep sea, he was a respected and prolific marine biologist who helped push the field forward.

1847

Joseph Hooker realizes that microscopic, planktonic diatoms can photosynthesize, and suggests that they play a similar role in the oceans as that of plants on land.

1855

Matthew Fontaine Maury publishes the first edition of *The Physical Geography of the Sea*, considered by many to be the first oceanography textbook.

1868–69

Charles Wyville Thomson leads dredging trips on HMS *Lightning* and HMS *Porcupine* that greatly expand scientific knowledge of the deep sea, finding representatives of most major animal groups at 650 fathoms (1,200 m) depth and demonstrating that temperatures in the deep sea vary considerably from place to place, suggesting complex circulation patterns.

1872–76

HMS *Challenger* circumnavigates the globe with the goal to investigate "everything about the sea." Sponsored by the British Royal Navy, it makes stops in all major oceans except the Arctic. The resulting fifty-volume *Challenger Report* is a classic of oceanographic science.

1882

The U.S. Fisheries Commission steamer *Albatross*, the first purpose-built oceanographic research vessel, takes to the seas. Research cruises led by scientist Alexander Agassiz in 1899 and 1904 greatly expand geophysical understanding of the Pacific Ocean.

1893–96

Norwegian Fridtjof Nansen floats around in Arctic sea ice in the *Fram*, a ship he designed to withstand the crushing pressure of the ice. The *Fram* is later used by Roald Amundsen as part of his successful bid to become the first person to reach the South Pole.

Milestones in Ocean Science (continued)

1902

The International Council for the Exploration of the Sea (ICES), the oldest such organization focused on marine and fisheries science, is established in Copenhagen with eight member nations.

1912

Alfred Wegener proposes the theory of continental drift. Although Wegener was not the first to propose that continents move across the surface of the globe, he assembled a comprehensive analysis of geographical, geological, bathymetric, and biological observations in support of the idea. Although Wegener did not live to see his theory widely accepted, he set the stage for the formulation of the plate tectonic theory in the 1960s.

1912

The *Titanic* sinks after hitting an iceberg, generating strong interest in the development of ways to detect objects in the water in front of moving ships. Two years later Submarine Signal Corporation's Reginald Fessenden successfully demonstrates a device that can reflect sound signals simultaneously from icebergs and from the seafloor, signaling the beginning of acoustic exploration of the seas.

1925–27

The German *Meteor* expedition combines biological studies of the Atlantic with the first wide-scale use of echo-sounding technology to map the bottom. Data from this expedition proves that the Mid-Atlantic Ridge runs the length of the Atlantic Ocean.

1930

William Beebe and Otis Barton make the first scientifically motivated deep-sea dive, using a bathysphere, a steel-walled spherical vessel designed by Barton.

1941

Rachel Carson publishes *Under the Sea Wind*, the first of three influential books she is to write about life in the ocean. Although best known for her 1962 book *Silent Spring*, sometimes called the inspiration for the environmental movement, Carson was a great marine ecologist as well.

1943

The U.S. Navy creates an Oceanographic Unit within the Navy Hydrographic Office. Headed by Dr. Mary Sears, the Oceanographic Unit rapidly grows in importance, eventually becoming the Naval Oceanographic Office.

1943

Jacques-Yves Cousteau and Emile Gagnan invent the Aqualung, the precursor to modern scuba equipment. Cousteau goes on to become an ardent spokesman for the oceans, producing several award-winning films about undersea life and receiving numerous awards for his conservation work.

1947

Maurice Ewing of Columbia University in New York leads a two-month research cruise on the RV *Atlantis*; he subsequently persuades the university to establish a separate research facility (now called Lamont-Doherty Earth Observatory) dedicated to geophysical oceanography.

1955

Bruce Hamon and Neil Brown develop the Conductivity-Temperature-Depth (CTD) meter, one of the most basic tools of modern oceanography. CTD meters allow scientists to measure how salinity (as indicated by conductivity) and temperature change with depth.

1957

Marie Tharp and Bruce Heezen publish a map of the North Atlantic seafloor. It is the first attempt to systematically map out what the world would look like if the oceans were removed.

1957

Roger Revelle and Hans Seuss publish a scientific paper that suggests the oceans cannot absorb all the carbon dioxide released by human industrial activity, and propose a link between human carbon-dioxide emissions and an increasingly strong greenhouse effect.

1960

Jacques Piccard and Donald Walsh set the record for the deepest manned dive, taking the bathyscaphe *Trieste* to the bottom of Marianna Trench, 35,800 feet (10,911 m) below the surface.

Testing the Fessenden oscillator

1961–62

Robert Dietz and Harry Hess suggest a mechanism for seafloor spreading that becomes the basis for the theory of plate tectonics.

1968

First cruise of the Deep-Sea Drilling Project (precursor to the Ocean Drilling Program) to investigate the evolution of ocean basins by drilling into ocean sediment.

1969

Robert Paine introduces the concept of a keystone predator, based on studies of the effect of a particular species of sea star on rocky intertidal communities on the Washington State coast.

1971–80

International Decade of Ocean Exploration (IDOE). The goal of IDOE was to create a cooperative, international research effort that would significantly increase our knowledge and understanding of the world's oceans. IODE's legacy continues today with the Global Ocean Observing System (GOOS).

1972

In recognition of the need to better understand and protect marine resources in the United States, the U.S. Marine Sanctuary Program and the National Estuarine Research Reserve System are established. There are now 14 federal marine sanctuaries and 27 federal estuarine research reserves in the United States.

1977

Deep-sea hydrothermal vents are discovered along the Galapagos Rift by scientists in the submersible *Alvin*. The abundance and diversity of life at these vents stuns the scientific world.

1978

NASA launches SeaSat, the first satellite designed expressly for oceanographic purposes.

1985

Sylvia Earle sets the record for solo diving in a submersible, reaching a depth of 3,280 feet (1,000 m). Earle, also called Her Deepness, pioneers research on open-ocean and deep-sea ecosystems.

1989

The oil tanker *Exxon Valdez* runs aground on the Bligh Reef in Prince William Sound, Alaska. Although far from the biggest oil spill in history, the *Valdez* oil spill is notable for its remote location and the coordinated efforts to understand the effects of the spill.

1990

Evelyn Fields becomes the first woman and the first African-American to serve as a commanding officer on a NOAA ship.

1992

TOPEX/Poseidon, a satellite that measures ocean surface height, currents, waves, and tides over most of the ice-free oceans every 10 days, is launched. The satellite is a joint venture between NASA and France's Centre National d'Études Spatiales (CNES). Following a malfunction, it is turned off in January 2006.

1995

The U.S. government declassifies Geosat satellite radar altimetry data, which allows worldwide seafloor mapping from space.

1998

Following record-high ocean temperatures resulting from a strong El Niño, massive bleaching and death of coral reefs occurs around the globe, with more than 16 percent of shallow-water reefs killed. Another severe bleaching event affects reefs in 2002, and scientists predict that such events will become increasingly common as global warming continues.

2003

The *Sorcerer II* expedition sets out to circumnavigate the globe studying marine microbes. The team discovers 1,800 new species and 1.2 million new genes in the Sargasso Sea alone.

2004

A deep-sea volcanic eruption is captured on video for the first time.

2005

Several hurricane records are set during this year. There are 15 Atlantic hurricanes, three more than the previous record set in 1969. Four of these belong to the strongest category of storms—category 5. Hurricane Wilma sets the record for strongest hurricane ever, prompting some to suggest the need for a new category of super-hurricanes.

Satellite image of Hurricane Katrina

Ocean Zones and Currents

ZONES OF MARINE ENVIRONMENTS

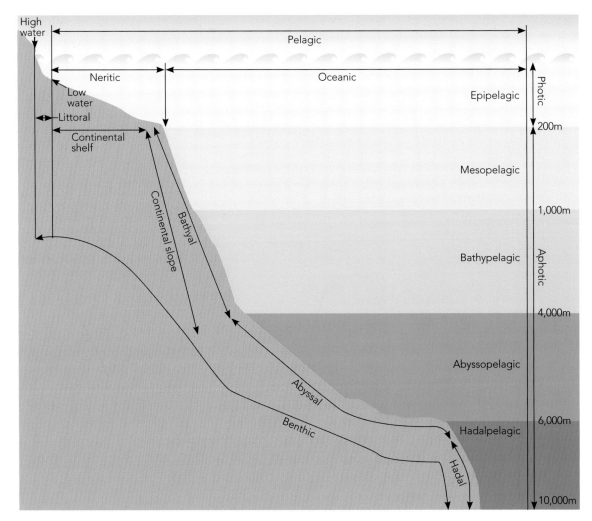

Photic: receives enough light for photosynthesis

Aphotic: receives too little sunlight for photosynthesis

Pelagic: the water column

Oceanic: open ocean beyond continental shelf

Neritic: shallow water, from edge of ocean to edge of continental shelf

Benthic: the seafloor

Littoral: intertidal from mean high to mean low water

Sublittoral: the seafloor from mean low water to the edge of the continental shelf

Bathyal: the seafloor on the continental slope and continental rise

Abyssal: the seafloor from the edge of the continental rise to about 6,000 meters

Hadal: the deepest zone

OCEAN SURFACE CURRENTS

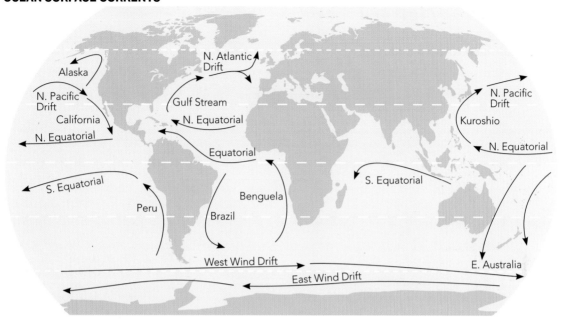

Major Surface Currents

Major surface currents are driven by the wind, gravity, the Coriolis effect, and the shapes of continents and ocean basins. The map shown above identifies the currents. The chart shows global wind patterns. The relationship between the two is described below:

North and South Equatorial Currents are driven by the trade winds;

West Wind, North Atlantic, and North Pacific Drifts are driven by the westerlies;

East Wind Drift is driven by the polar easterlies;

Peru, California, Alaska, Kuroshio, Canary, East Australian, Benguela, and Brazil Currents and the Gulf Stream are not driven directly by the winds, but rather by the Coriolis effect and pressure gradients created by east-west currents.

GLOBAL CIRCULATION

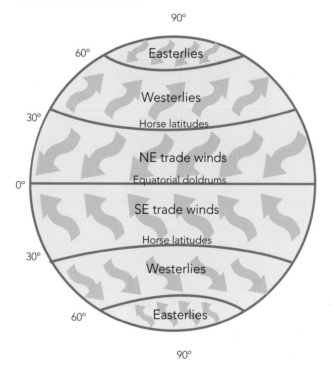

Geology and the Oceans

PLATE BOUNDARIES

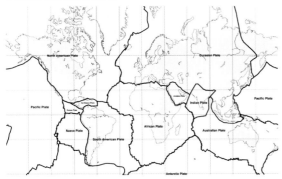

Locations of major tectonic plates. Most plate boundaries are generally agreed upon, but others are less certain.

EARTHQUAKES

Location and depth of earthquakes occurring between 1990 and 1996. Depth is indicated by color: Red is shallowest, followed by orange, green, and blue. The deepest quakes occur 200 to 400 miles (300–700 km) below the surface.

SEAFLOOR AGE

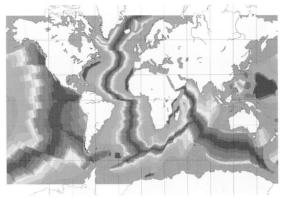

Color indicates seafloor age as determined by radiometric dating, ranging from new (darkest red) to around 50 million years old (yellow) to 180 million years old (darkest blue).

VOLCANIC ERUPTIONS

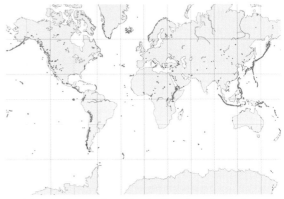

Red dots mark currently or historically active volcanic features. Most are volcanoes, but hot springs, geysers, and other such features are also included.

The theory of plate tectonics states that the surface of the Earth is made up of a number of distinct plates, the size and position of which change gradually over time. This theory was initially dismissed but within the last half century has become widely accepted, revolutionizing our understanding of the geological world. A variety of evidence led geologists to the theory of plate tectonics. Notice that volcanic features and earthquakes are most concentrated along certain plate boundaries. These are subduction zones, where one plate is thrust under another. Divergent plate boundaries, where two plates are moving away from each other, have primarily shallow earthquakes, while convergence zones (where plates collide) may have very deep quakes. As one would expect, the seafloor is youngest at spreading centers and gets progressively older as it gets farther away.

SEAFLOOR MAP

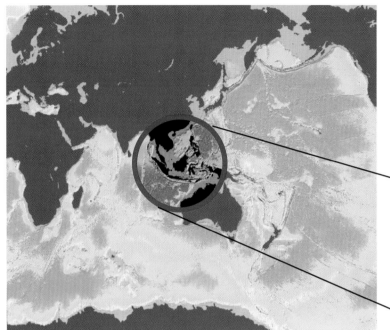

This bathymetric map was created using a technique called satellite altimetry. The topography of the seafloor is reflected in small differences in the height of the ocean's surface. While these dips and bulges are not visible to the naked eye, they can be accurately measured with specialized equipment on a satellite. The inset (below) shows the area around Sumatra. The yellow star marks the epicenter of the December 2004 tsunami-causing earthquake.

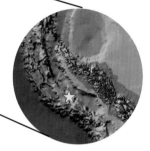

Matthew Fontaine Maury published the earliest known map of seafloor bathymetry (depth) in 1853 using a handful of depth soundings. Even today, the depth of the vast majority of the seafloor has yet to be directly measured, although satellite data allows scientists to predict the topography of the seafloor fairly accurately. Bathymetric maps provide further support for the theory of plate tectonics. Oceanic plate boundaries are marked by mid-ocean ridges. Where the edge of a continent lies along a subduction zone, depth drops off steeply, while other continental margins have wide, shallow shelves.

These false-color images show how two major earthquakes in late 2004 and early 2005 reshaped the coastline of the island of Nias, near Sumatra. Coral reefs that bordered shallow lagoons in 2000 (left) were high and dry by April 2005 (right).

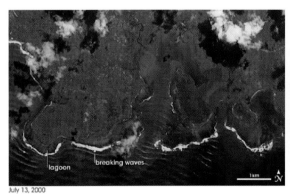

lagoon
breaking waves
1 km
July 13, 2000

exposed coral (uplifted)
April 6, 2005

The Diversity of Life

THE TREE OF LIFE

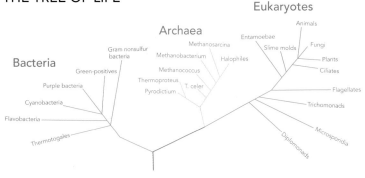

This tree represents a hypothesis about how living things are related. It is based primarily on genetic sequences and cell structures from hundreds of different species and is the simplest interpretation of existing evidence.

Aristotle is often credited with creating the first classification of life. Over 2,000 years ago, he categorized living things as either plants or animals and further identified animals as air, land, and water dwellers. If existing hunter-gatherers are any guide, though, humans have been classifying the world around them for much longer, primarily based on what they could eat, what had medicinal uses, and other utilitarian concerns.

Today, classification is often based on evolutionary relationships. Using available evidence for how groups of organisms might be related to each other, scientists draw the equivalent of a family tree, called a phylogenetic tree. The groups may be phyla, species, or any other taxonomic group. At the bottom of the tree, a single node represents the last common ancestor of all organisms on the tree. The branches extending from the trunk of the tree show where a single group splits into two or more sections, as when chimpanzees and humans arose from a single common ancestor. The tip of each branch represents one of the groups of organisms under study. Like family trees, they represent an ancestor (although family trees start with two ancestors, a male and a female) and all its descendants. Family trees usually have the ancestors on top and the descendants on the bottom, while phylogenetic trees usually have the ancestor on the bottom and the descendants on the top. As new organisms are discovered, or new morphological, cellular, and genetic information comes to light, phylogenetic trees may need to be redrawn.

The three major divisions of life are called domains—domain Bacteria, domain Archaea, and domain Eukaryota—and significant groups within these domains are called kingdoms or phyla. Early in the history of life, genes and even whole organisms from one domain may have been incorporated into organisms from another. Cellular and molecular evidence suggests that parts of our own cells, called mitochondria, began as bacteria that took up residence inside other cells, eventually losing the ability to live independently. Since early Earth had no land, the first stirring of life almost certainly happened in the sea. The tree of life shows how much those first marine life forms have varied—how small and simple marine cells gave rise to the amazing diversity that exists today. Early life very likely split into domains while still residing in the oceans. The point on the tree diagram where all the lines connect represents some ancestral marine organism that gave rise to a multitude of "children" on Earth.

DEFINING DIVERSITY

Biodiversity is often defined as the total number of species in an area. Since the species concept works poorly or not at all for bacteria and archaea, scientists differ on how diversity should be defined. Some advocate genetic diversity, while others argue that functional diversity—how organisms get nutrition, reproduce, defend themselves, and so on—matters most. However you define diversity, the oceans are full of it.

Ways of Getting Nutrition

Living things need carbon and energy. Heterotrophs get carbon from organic matter (other living or once-living things, such as meat or sugar): *hetero* is Greek for other, and *trophy* for feeding. Autotrophs get carbon from carbon dioxide (CO_2): *auto* is Greek for self. Energy sources include light, inorganic material (such as rocks), or organic matter. Photoautotrophs (such as plants and algae) receive energy from light and carbon from CO_2. Photoheterotrophs (such as purple non-sulfur bacteria) get energy from light and carbon from organic matter. Chemoautotrophs (such as deep-sea vent bacteria) get energy from inorganic material and carbon from CO_2. Chemoheterotrophs (such as animals) receive energy from organic or inorganic sources and carbon from organic matter. All living things fall into one or more of these categories.

Proteobacteria

This is the most ubiquitous and abundant bacterial phylum in the water, and includes the bacteria that cause cholera and are key members of deep-sea vent communities.

Cyanobacteria

Often misleadingly referred to as blue-green algae, they are not algae at all. Often forming clumps or chains, they are important sources of nitrogen.

Domain Bacteria

Bacteria are single-celled and prokaryotic, meaning that their DNA is not contained in a membrane-bound nucleus. There are four basic forms—rods, spherical cocci, comma-shaped vibrio, and helices—but this group is more physiologically diverse than plants and animals combined, covering all four approaches to getting nutrition (plants and animals between them use just two!). There are 20 known phyla of bacteria, but the true number is undoubtedly higher.

Green sulfur bacteria

This group includes some of the most efficient photosynthesizers known, living 260 feet (80 m) below the surface but still using sunlight as an energy source. All consume sulfur compounds produced by decay and hydrothermal activity. One species lives near black smoke vents.

Spirochaetes

This group includes the bacteria that cause Lyme disease and syphilis, as well as species that live freely in mud and sediment. They are thin, spiral-shaped, and flexible.

Salt-lovers (halophiles)

Halophiles thrive in high-salt environments like the Dead Sea and evaporating ponds of seawater, often requiring salt concentrations ten times higher than normal seawater.

Heat-lovers (thermophiles)

Archaen thermophiles grow best between 176°F and 212°F (80–100°C), the boiling point of water at sea level. In contrast, no known eukaryote can survive above 140°F (60°C). They are common in hydrothermal vents, often preferring low pH as well as high temperatures.

Domain Archaea

Despite a superficial resemblance to bacteria—both are single-celled prokaryotes—archaea are no more closely related to bacteria than to humans. Our understanding of this group is in its infancy, and archaea are often classified by where they live rather than to whom they are related.

Methanogens

These archaea produce methane, and die if exposed to gaseous oxygen. They are common in oxygen-poor marine sediment. Most natural gas deposits mined by energy companies were created by methanogens.

Cold-lovers (psychrophiles)

These archaea live only in permanently cold areas such as Antarctic sea ice. They can survive and grow at temperatures below freezing.

Domain Eukaryota

Eukaryotes include fungi, protists, algae, true plants, and animals, including humans. The defining feature of a eukaryote is a membrane-bound cell nucleus. In bacteria and archaea, the chromosomes are loose in the cytoplasm.

Fungi

While the idea of marine mushrooms and mold might seem odd, all major groups of fungi can be found in the marine environment. Important

decomposers, they are particularly common in coastal areas with a lot of detritus, such as mangrove forests and salt marshes. Marine fungi may also be parasites, and fungi in the genus *Aspergillus* can cause widespread mortality of sea fans and some species of reef-building corals.

Algae

The organisms generally referred to as algae represent three evolutionarily distinct groups. Brown algae fall in the same group as mildew, while red algae are a completely distinct group no more closely related to brown algae than to several other groups of protists, and green algae are most closely related to plants.

Rhodophyta (red algae):

Red algae derive their color from a pigment that reflects red light and absorbs blue. Blue light travels deeper into water than other light, so red algae can often live at greater depths than other algae.

Green algae:

Some scientists put green algae in the kingdom Plantae, and many freshwater green algae are more closely related to plants than to other green algae.

Diatoms:

Unicellular algae with rigid silica cell walls that both protect the cell and allow sunlight to pass through for photosynthesis.

Alveolates:

Members of this single-celled group may be heterotrophic or photosynthetic, and all share the characteristic of a system of sacs located beneath their cell membranes. They include foraminifera, ciliates, and dinoflagellates, famous for their bioluminescence and their role in red tides and fish kills.

Sea grasses:

There are about 60 species of sea grasses, a group distinct from the grasses that grow on land. They can form extensive, highly productive subtidal or intertidal meadows.

Coccolithophores:

Single-celled algae encased in complex scales that are usually made of calcium carbonate.

Chromista (brown algae):

Brown algae range from microscopic, single-celled organisms to kelps over 150 feet (46 m) long. Most are photosynthetic, but water molds—once considered fungi—usually get their nutrients from decaying plants and animals.

Kelps:

Large, multicellular algae. They may form extensive offshore "forests" that protect the coast from erosion and provide a three-dimensional habitat for numerous species including crabs, fish, and sea otters.

Protists

The term protist is just a fancy way of saying "a eukaryote that isn't a plant, animal, or fungus." Though considered a single kingdom now, protists are more diverse than animals, plants, or fungi, and so will probably be broken into several kingdoms as knowledge of them increases.

Radiolarians:

These common planktonic organisms have glasslike, intricate skeletons made of silica. Some species are predatory, and others harbor symbiotic algae.

Vascular, or true, plants

Unlike green algae, moss, and other "simple" plants, vascular plants possess specialized tissues for transporting water, including roots, stems, and leaves. Only a few groups of plants are truly marine, meaning they are found exclusively in marine habitats.

Mangroves:

The term mangrove refers to several different groups of trees in tropical and subtropical coastal marine environments. Aerial roots and physiological mechanisms for excluding or excreting salt make this vegetation highly adaptable to high salinity and low oxygen conditions.

Salt marsh plants:

Salt marshes occur in low-energy intertidal areas around the world, and they hold the greatest diversity of marine plants. Several true grasses are found here, as well as rushes and sedges (in the same botanical order as true grasses), sea lavender (in the same botanical order as carnations and cacti), and plantains (in the same botanical order as mint and verbena).

Kingdom Animalia

Our understanding of the evolutionary relationships among animals, like most scientific ideas, is revised as new evidence comes to light. This diagram represents a new and by no means universally accepted hypothesis about animal relationships. Animals are divided into those with bilateral symmetry, where the right and left sides are roughly mirror images of each other, and those without. The Bilateria are further split into three major groups: the Ecdysozoa, or molting animals; the Lophotrochozoa, which share distinct feeding or larval types; and the Deuterostomia, defined by developmental patterns. About half of the animal phyla are shown here.

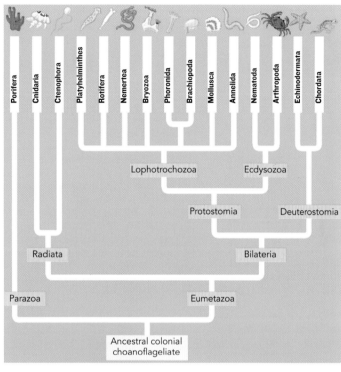

Porifera
Cnidaria
Ctenophora
Platyhelminthes
Rotifera
Nemertea
Bryozoa
Phoronida
Brachiopoda
Mollusca
Annelida
Nematoda
Arthropoda
Echinodermata
Chordata

Lophotrochozoa
Ecdysozoa
Protostomia
Deuterostomia
Radiata
Bilateria
Parazoa
Eumetazoa
Ancestral colonial choanoflageliate

Sponges

Cnidarians

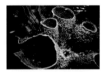

Ctenophores

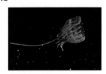

Arthropods

Nematodes

Annelids

Mollusks

Brachiopods

Echinoderms

Tunicates

Vertebrates

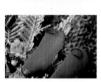

Tools of the Trade

COLLECTING SAMPLES

Collecting samples of water, rock, sediment, ice, and living organisms for examination in laboratories allows researchers to make measurements that might not be possible in the field and to work with organisms in controlled environments.

Nets

Plankton nets are used to capture microscopic organisms from the water column. They may be pulled through the water by hand or winch, or held on a dock or boat while water is pumped through them.

Scientists collect larger organisms with nets similar to those used by fishermen. Seines are vertical nets for capturing animals near the surface, and dredges and trawls are heavy nets or bags dragged from boats for collecting animals from the bottom. Dip nets, small nets on a pole, are used to collect individual animals near the surface.

Plankton nets

Bottles

The simplest bottles are used to scoop a sample of water by hand. Sterile bottles are used for work with bacteria or viruses to ensure that any such microscopic entities found come from the sample and not from contamination in the lab.

To sample deeper water, bottles are hung from a cable or line and operated remotely. Van Dorn bottles are open-ended cylinders with lids on both ends. The lids are held open by a catch, and when it is time to take a sample, a metal weight called a messenger drops down the cable to trigger the bottle to close. Niskin bottles are similar, but they are equipped with thermometers that record the temperature of the water sample when it was taken.

Niskin bottles

Sediment Samplers

Grabs are hinged mechanical jaws that collect samples from the ocean floor, disrupting its structure. Ekman grabs are made of brass, triggered by a messenger, and work best for muddy or sandy bottoms. The much heavier Petersen grab is used for hard substrates, and it can be triggered with a messenger or by its own weight.

Corers provide a cylindrical sample of the bottom that keeps the vertical structure intact, giving researchers a history that may stretch back millions of years. Gravity corers are driven into the seafloor by their own weight and can collect samples several meters long. Piston corers can go even deeper, up to 90 feet (30 m). Both these corers are best used for softer substrates. To get samples of hard rock bottom in the open ocean, scientists use special deep-sea drilling ships capable of collecting cores thousands of feet long in the deepest parts of the ocean.

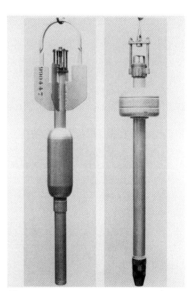

Corers

COUNTING AND MEASURING

Counting the number, diversity, and density of organisms and monitoring their movements enriches the understanding of marine ecosystems. Equally critical is measuring how the physical and chemical environment in which these organisms live changes over time and locale.

How Many

Since counting all the organisms along a stretch of shoreline or seafloor is not always practical, scientists often use transects and quadrats to get accurate estimates of the number and density of organisms in a given area. Transects are lines along which counts may be made, while quadrats allow counting within an area rather than along a line.

Mark-recapture estimates of population size involve marking individual animals with dye, collars, bands, or other devices, and seeing what percent of animals captured at a later time are marked.

Quantifying planktonic organisms requires a microscope and some sort of counting chamber. Hemocytometers and Sedgwick counting chambers are good for small organisms and small sample volumes. For larger organisms and samples, Bogoroff trays and plankton settlement chambers are preferred.

Banded bird

On the Move

Small, battery-powered tags that transmit radio, audio, or satellite signals are useful for tracking animal movements over large distances. Tags may simply emit signals helping researchers locate animals, or they may store data on the animal's movements, which are transmitted to researchers at a later time. Tags may be glued or strapped to an animal, but these work only for animals large enough not to be harmed by the weight or size of the radio tag.

Animal movement can also be tracked by equipping animals with bands or tags that provide information on when and where they were first caught.

CTD meter

Chemistry

Perhaps the most ubiquitous piece of oceanographic equipment is the conductivity-temperature-depth meter, or CTD. CTDs can measure salinity using the water's ability to conduct electricity, temperature, and depth from the surface down to the bottom of the ocean. Salinity can also be measured using hand-held hydrometers or salinometers.

A common tool for measuring water clarity is the Secchi disk, a white or black-and-white disk that is slowly lowered into the water until it can no longer be seen from above. Clarity can also be measured using transmissometers and optical backscatter meters. These tools measure how much light is transmitted and scattered by water by shining light of a particular wavelength through a known volume of water and measuring how much light is absorbed or scattered back toward the light source.

A number of sensors and test kits also exist to test pH, dissolved oxygen concentration, and other chemical characteristics of water. Some are cheap, simple, and less accurate, while others are designed to function with high levels of accuracy at great depths.

Currents

Although a number of high-tech options for measuring currents are now available, simple drift cards or floats are still common. Each card or float is printed with a unique identification number and an address or phone number to be used by whoever finds the card or float. This allows oceanographers to track major current patterns.

Currents can also be measured in a particular location using an Acoustic Doppler Current Profiler, which uses the way movement affects sound to measure water speed and direction.

Acoustic Doppler Current Profiler

Tools of the Trade (continued)

VISUALIZING THE UNSEEN

Most of the ocean is not visible to the naked eye, so oceanographers have had to come up with clever ways of figuring out basic things like how deep the oceans are and what the bottom looks like.

Seafloor Depth and Character

The depth and character of the ocean floor can be measured with sound-based techniques using different frequencies for different purposes. Side-scan sonar provides a picture of the seafloor and objects resting there. Echo sounders measure water depth. Sub-bottom profiling, or seismic reflection, provides information on what lies below the surface, going as deep as 160 feet (50 m). Depending on the sound frequency used, it generates images of geological layering and structure below the surface of the seafloor, objects buried in the sediment, or even the density structure of the water column.

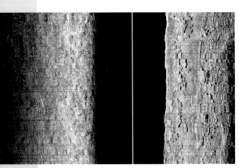

Sonar scan

Satellites

Satellite imagery has made oceanography truly global. Satellites carrying Advanced Very High Resolution Radiometer (AVHRR) sensors measure sea surface temperature. The Sea-viewing Wide Field-of-view Sensor (SeaWiFS)

measures chlorophyll, allowing scientists to track primary productivity across time and space. Wind speed and direction are measured by SeaWinds, a microwave scatterometer. Spectacular images with a resolution as small as 820 feet (250 m) are generated by Moderate Resolution Imaging Spectroradiometers (MODIS), 36-channel optical scanners currently carried on two satellites. Until its retirement in January 2006, the TOPEX/Poseidon satellite carried a radar altimeter that measured sea surface height. The Jason satellite has replaced TOPEX/Poseidon.

Cameras

The Video Plankton Recorder uses an underwater microscope to capture video footage of organisms ranging in size from a thousandth of an inch up to about an inch. This allows scientists to see the composition of the plankton before it might get damaged or destroyed by traditional net- or bottle-collecting techniques.

Submersibles

A number of submersibles have taken scientists to the bottom of the ocean over the years. The best-known submersible is the *Alvin*, a three-person vehicle built in 1964 by the Woods Hole Oceanographic Institution. Initially able to dive safely to just 6,000 feet (1,830 m), *Alvin* was retrofitted in 1973 to reach depths of over 13,000 feet (4,000 m). The *Alvin* enabled scientists to discover hydrothermal vents and make a firsthand inspection of the wreck of the *Titanic*. Other, more sophisticated submersibles have since been developed. Subs allow scientists to actually see the bottom of the sea and to collect samples and photographs of what they find there.

BOATS AND ROBOTS

Ever since the *Challenger* made its debut voyage in 1872, boats designed specifically for oceanographic research have been deepening our understanding of how oceans work. These boats are expensive to build and operate, so oceanographers have also designed and deployed a number of autonomous underwater vehicles, or AUVs, which can travel long distances at many depths, collecting data for months or years without need for human input.

Boats

One of modern oceanography's most prolific research vessels is the RV *Atlantis*, operated by the Woods Hole Oceanographic Institution. The mother ship for the submersible *Alvin*, the remotely operated *Jason*, and two towed vehicles, it is equipped with

Alvin

sophisticated navigation, bottom mapping, and satellite communication capabilities.

Some ships, like the JOIDES *Resolution*, are designed specifically to support deep-sea drilling. Originally used for oil exploration, the *Resolution* was converted to a scientific research vessel in 1978. Twelve computer-controlled thrusters keep the ship precisely positioned during multiday drilling operations.

Icebreakers are boats built to withstand and cut through sea ice, allowing scientists to access locations that would otherwise be unavailable.

A different sort of vessel is the research platform FLIP. Built to provide a stable place for mid-ocean studies, the FLIP is towed to sea in a horizontal position, then filled almost completely with water. Just 55 of FLIP's 355 feet (108 m) stick above the surface. This makes it extremely stable, even in large waves.

Autonomous Underwater Vehicles

RAFOS floats create high-resolution maps of subsurface currents over large areas, using sound from fixed acoustic beacons to track their speed and location. An earlier type of float, SOFAR (sound fixing and ranging), tracked currents by sending out acoustic signals that were picked up by fixed listening stations. Since RAFOS works in the opposite way, its name is SOFAR spelled backward.

RAFOS floats are glass tubes about 6.5 feet (2 m) long that weigh 22 pounds (10 kg) each and are ballasted to float at a particular depth using stainless steel weights attached to the outside. Sensors measure pressure, temperature, and dissolved oxygen as well as current. An internal microprocessor stores data and, at a preprogrammed time, signals the float to drop its ballast and float to the surface. Once on the surface, the float beams data

to two satellites, which transmit the information to ground receiving stations that in turn send it over the Internet to the appropriate people. The floats can be programmed for missions of up to two years.

Spray Glider

Simple, cheap, and effective, spray gliders navigate a preprogrammed underwater course using a small onboard computer. They can carry a variety of sensors that measure things like temperature, salinity, turbidity, and currents. At the start and end of a dive, which may be up to six months long, a glider rolls onto its side so its GPS antenna can measure the location. The glider communicates with researchers via satellite.

Spray gliders are 6.5 feet (2 m) long with a wingspan of almost 4 feet (1.2 m). Except for a plastic tail section housing sensors and a steering device, the glider's body is made of aluminum.

Floating Instrument Platform (FLIP)

CHAPTER 7

WAVES AND TIDES

Left: Waves pound the Cliffs of Moher on the Atlantic coast of Ireland. Tides and waves shape the world's coastlines. Top and bottom: The power of waves is both enjoyed and feared by people living near oceans. Storms at sea can lead to destructive wave activity including tsunami waves, which are the result of underwater earthquakes and volcanic eruptions.

Two of the most familiar features of the ocean are tides and waves. Waves rise and fall in response to wind, earthquakes, and other forces. They can be immensely destructive, like the tsunami that struck Asia in 2004. They can also provide endless pleasure for surfers. And while the most visible waves occur on the surface of the water, waves also occur underneath the surface, forming where layers of different densities meet. Because they are harder to observe, these internal waves remain much less understood.

Tides, the biggest waves of all, rise and fall on a daily basis. While the "ideal" tides predicted by Newton were driven solely by the gravitational pull of the Sun and Moon, tides in the real world respond to the depth and shape of ocean basins, interactions with continents and islands, and myriad other forces. The difference between high and low tide varies from almost nothing to more than 30 feet (10 m). In some parts of the world, currents generated by the tides can provide an ample source of renewable energy.

The Anatomy of a Wave

Big or small, curling or frothy, waves are a central element of the seashore. Waves ranging from the smallest surface waves to the Earth's greatest waves, the tides, share some common traits. Three basic characteristics of a wave are its height, length, and period. The distance between a wave's highest and lowest points, known as its crest and trough, defines its height. Its length is simply the distance between two crests, and the time it takes for two successive crests to pass the same point determines its period. How often a crest passes a fixed point is the wave's frequency. The size of a wave is influenced by how much energy goes into making it. For instance, a strong wind has more energy than a light wind, so it will generate bigger waves. Waves are also bigger when the wind has more time to transfer energy to them, as happens when the wind travels unimpeded over a large expanse of water. The distance the wind travels is known as the fetch.

MAKING AND BREAKING WAVES

Waves are often classified by their cause—called the disturbing or generating force—and the force that returns the water to equilibrium, known as the restoring force. The most common disturbing force is wind, but earthquakes, landslides, and other forces generate waves as well. For very small waves, the surface tension created by the attraction of water molecules to one another is enough to restore equilibrium. These waves are known as capillary waves and are generally short-lived. For larger waves such as tsunamis or breakers, gravity rather than surface tension is the restoring force, so they are called gravity waves. As waves move into shallow water, they slow down. Although the period remains unchanged, the wave height increases. Friction with the seafloor causes the bottom of the waves to slow more quickly than the top, and the front of the wave becomes steeper and steeper. Eventually it falls forward, or breaks, dissipating the wave's energy.

MOVEMENT INSIDE THE WAVE

Although waves and the energy they contain travel forward through the water, individual particles within the waves may

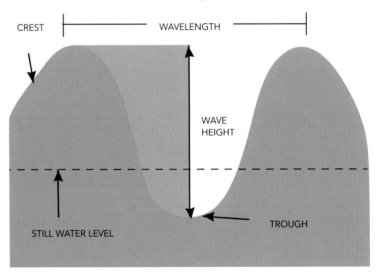

Above: Wave anatomy. The crest is the highest point on a wave and the trough is the lowest. Wavelength is the distance between the crests or troughs of two consecutive waves, and wave height is the difference in elevation between the crest and trough. Top left: A breaker occurs when a wave's height exceeds its vertical stability, causing it to tumble forward, or break, as it approaches shore.

As waves pass by an object floating on the water the object will move forward as the wave approaches, and then back as it passes. This movement causes the object to bob up and down on the surface.

not. An object floating on the water moves forward as a wave approaches, and then backward as the wave passes. Waves pass by boats floating on the surface rather than carrying the boats with them. This circular motion creates the bobbing of floating objects, moving up and down, backward and forward. In reality, the motion of objects in or on a wave is not perfectly circular, and some forward progress *is* made. This is particularly true when the relationship between water depth and wave height is such that wave motion is influenced by friction with the seafloor. Any wave traveling through water whose depth is less than half the wavelength is considered a shallow-water wave, regardless of the absolute depth of the water.

The USS Trinity, *which was the same type of ship as the USS* Ramapo. *These fleet oilers provided fuel, petroleum products, and some ammunition to battleships during World War II.*

THE RAMAPO MEASURES SOME BIG WAVES

Sailors, like anyone else, may be prone to exaggeration. The crew of the navy oiler USS *Ramapo*, however, could honestly say they had survived the biggest wave measured by humans. In early February 1933, the *Ramapo* was traveling from the Philippines to California when it was caught in a huge storm. The storm lasted for seven days and stretched across the entire ocean, generating tremendous waves. When a particularly big wave came by, an officer on deck noticed that he could see the crest of the wave straight across the foremast. Making some simple calculations, he put the wave's height at 112 feet (34 m), about the height of an 11-story building.

FREE AND FORCED

Most waves are progressive wind waves. That is, they are created by wind pushing on them from a particular direction, and they progress across the water in the direction in which the wind originally pushed them. While waves remain directly under the influence of the wind or other generating forces, they are called forced waves; however, waves may continue moving across the ocean when the wind or other generating force has stopped. These waves are called free waves, moving at speeds determined by their period and wavelength rather than by the wind. As waves move away from the center of a storm, for instance, longer waves travel faster than shorter ones. This creates sets of waves of uniform wavelength, called swells. Swells can travel across the world: Some have been tracked moving from the Antarctic to the Arctic.

Bouncing off the Walls

With winds, currents, and coastlines all influencing wave formation and direction, it is inevitable that groups of waves with different sizes and orientations meet. What then? In the region where waves intersect, the water moves in complex patterns. Two crests meet and create a crest that is twice the height of either wave alone. A crest and a trough meet and cancel each other out. Still, some waves pass through this region and move on, their size and direction unchanged. When waves meet a solid object, a different pattern is created.

REFLECTION

If a wave hits a surface that is steep enough, it bounces off it, reflecting back the way light reflects off a mirror. If the wave is moving perpendicular to a wall (that is, the wave front is parallel to the wall), the wave will reverse direction when it hits the wall. If the wave is moving at an angle to the shore, it reflects back at an angle equal to but 90 degrees off its incoming angle. These intersections create complex interference patterns like those described earlier, complex seas that can be tricky for small boats to travel through. If the waves are bouncing off a curved surface, the patterns get even more complex. Depending on how the surface is curved, wave energy may be dispersed or enhanced.

REFRACTION

On shorelines with shallow slopes, waves slow and break. Unless the wave line is exactly parallel to an even shoreline, however, different parts of the line will slow at different times. These variations cause the wave line to bend, or refract, as it moves into shallower water. To understand why this happens, imagine a wave crest as a line of people walking in unison. If

Above: Cliffs shaped by erosion tower over the Indian Ocean on the southern coast of Australia. Top left: Rocky coastlines interfere with wave patterns, causing waves to reflect, or bounce back.

people at one end of the line slow down, people in the rest of the line will have to swing around somewhat to maintain the line. When one end of the wave line hits shallow water, it slows down because of friction with the bottom, and the rest of the wave line bends to become more parallel to the shoreline.

Refraction gets more complex if the shoreline has bays and headlands. Wave lines hit the headland first and wrap around it. This concentrates wave energy on the headland, creating an area of high wave energy. The center of a bay is usually deeper than the sides, so wave crests bend away from the center, creating an area of low wave energy. This difference in wave energy creates a difference in erosion strength—stronger on headlands, weaker in bays—that has the effect of straightening shorelines over time.

DIFFRACTION

Yet another phenomenon can change the direction and pattern of waves: diffraction. Diffraction is the bending of waves as they pass through narrow openings or around obstacles. When waves pass through an opening in a breakwater, for instance, they spread out, transmitting the original energy over a much larger area. If a breakwater has two or more openings, the interaction of diffracted waves produces interference patterns behind the breakwater. Waves can also diffract, or slow and bend, as they pass the end of a jetty or other barrier.

Point Udall, St. Croix, from the air. Wave refraction patterns can be seen in the shallow water to the left of the point.

A seiche washes through Canal Park in Duluth, Minnesota, on Lake Superior.

STANDING WAVES AND SEICHES

When two wave groups whose waves have similar frequencies are traveling in opposite directions and they meet, the result is a phenomenon called standing waves. Here the peaks and troughs intersect such that some points, called nodes, remain at a fixed height, while between the nodes peaks and troughs twice the size of the original wave go up and down. This gives the impression that the waves are not moving forward. When standing waves occur in an enclosed or partially enclosed body of water such as a lake or harbor, they are called seiches and can have dramatic effects. In 1995 in Lake Superior, North America, a seiche formed such that the water along one stretch of shoreline rose and fell by 3 feet (1 meter) in just 15 minutes. Boats that were floating one minute were high and dry the next.

Tsunamis

On December 26, 2004, a powerful wave swept across the Indian Ocean, killing more than 200,000 people and leaving millions homeless. Triggered by a huge earthquake centered near the Indonesian island of Sumatra, the tsunami caused death and destruction as far away as eastern Africa.

Tsunamis are sometimes called tidal waves, but they have little to do with tides. They are caused by the rapid vertical displacement of large volumes of water. Although earthquakes on the seafloor are the most common cause, tsunamis can also result from large landslides, volca-nic eruptions, and (although no one has ever seen this) meteorite impacts. Tsunamis generated by landslides or volcanic eruptions usually do not travel as far as those generated by earthquakes, although they can still be quite powerful. A rockslide in Lituya Bay, Alaska, in fact, caused the largest tsunami ever recorded, surging to a maximum height of 1,720 feet (520 m) up the moun-tainside. Once the wave reached the open ocean, however, it rapidly diminished in power.

LONG AND SHALLOW

The long wavelengths of tsu-namis, often 60 to 120 miles (100–200 km) long, mean that although they travel across the open ocean, they are actually shallow-water waves. That is, they interact with the ocean bottom. An underwater ridge could re-fract the waves, or they might be diffracted as they pass between seamounts. A typical tsunami is 3 feet (1 m) high in the open ocean. The Indian Ocean tsunami was just 2 feet (60 cm) high two hours after the quake. Still, tsunamis are quite fast and can travel at speeds greater than 500 miles per hour (800 km/h).

Like all waves, a tsunami becomes taller as it moves into shallower water. The water at

Left: Meulaboh, Indonesia, photographed in May of 2004. Meulaboh is located on the coast of Sumatra, roughly 93 miles (150 km) from the epicenter of the earthquake that generated the tsunami of December 26, 2004. Right: Meulaboh on January 7, 2005, after the tsunami caused erosion and the destruction of many of the town's buildings. Top left: In 1963, Kodiak, Alaska, experienced a tsunami that caused fishing boats to wash into town.

the front of the wave slows, and water from behind piles on top of it. Tsunamis rarely break the way normal waves do, however. In fact, it is not so much the height of a tsunami that makes it so destructive; it is the sheer volume of water it carries.

WARNING SIGNS

Just as water pulls back before small waves, it pulls back just prior to tsunamis. Because tsunamis are so large, the water level may drop 10 to 13 feet (3–4 m) as the tsunami gets close to land. Unwary beachgoers may be drawn out to the exposed area by curiosity or to collect stranded fish, but this is just the wrong thing to do. The tsunami brings water back in to shore quickly, and the water level may go from unusually low to 20 or 25 feet (6 or 8 m) above normal in just four or five minutes. After another few minutes, the water recedes, taking people, animals, and debris with it.

NEVER ALONE

Tsunamis are almost always part of a wave train, or group of waves traveling together across the ocean. While normal waves have periods on the order of several seconds, tsunamis may have periods ranging from 10 minutes to a few hours. This means that one tsunami may be followed by a second or even third large wave minutes or hours after the first wave hits. In December 2004, parts of southern Thailand were hit by a wave about 15 feet (4.6 m) high one hour after the first tsunami struck.

On April 1, 1946, a Pacific-wide tsunami affected Hawaii (above) and the Aleutian Islands. Waves 30 feet (9 m) high reached as far as French Polynesia. This disaster led to the establishment of a tsunami warning system.

TSUNAMI WARNING SYSTEMS

If people are warned of an approaching tsunami in time, they can escape. In 1946, a particularly strong tsunami originating in the Aleutian Islands inspired the creation of the Seismic Sea Wave Warning System, later renamed the Pacific Tsunami Warning System. Twenty-six countries monitor seismic and tidal stations around the Pacific Ocean. If a tsunami seems likely, warnings are sent out to government officials and the media, and to the public via marine radio frequencies. Although 15 of the 20 warnings issued since 1946 have been false alarms, the possibility for severe destruction makes such a system worthwhile. Because the Pacific-wide system takes an hour or so the generate warnings, many countries around the Pacific have regional warning systems that can generate warnings in 5 or 10 minutes. No such systems currently exist in the Atlantic and Indian oceans, where tsunamis are much less frequent. Had a warning system existed for the Indian Ocean, thousands of lives might have been saved in the 2004 tsunami. This knowledge has spurred an international effort to create global tsunami warning systems, with efforts particularly focused on the Indian Ocean and the Caribbean. Creating such systems involves the installation of sea-level gauges, enhanced seismographic networks, and deep-ocean pressure sensors, as well as the establishment of regional warning centers linked to disaster management systems.

Waves beneath the Waves

The waves with which we are most familiar are those that happen at the boundary of air and water. Nonetheless, waves can form at the boundary between any two layers of fluid with different densities: air and water, freshwater and salt water, or even layers of air with different densities. When waves form underwater, they are known as internal waves.

Several forces can generate internal waves, including storms, water traveling over an uneven surface, and a steep current gradient between the layers of water above and below the pycnocline. Internal waves are particularly common in areas where freshwater flows into the sea, such as at the mouth of large rivers or in fjords and estuaries. In some cases, internal waves may move parallel to surface waves. This is called the fast or surface mode. More commonly, internal waves travel in the slow or internal mode, where their peaks and troughs are offset from the peaks and troughs of surface waves, with the peaks of internal waves often underneath the troughs of surface waves.

BIG AND SLOW

For large waves, the restoring force is a product of gravity and the density difference between the two layers of fluid. Because the density difference between

Above: This image shows ocean characteristics of the Gulf of Mexico northeast of the Yucatan Peninsula. Internal waves are visible at the center. Top left: Oslo Fjord, Norway. A fjord is a deep, U-shaped valley with steep walls that was formed by glacial erosion and filled with seawater. This geometry makes internal waves common in fjords where fresh and salt waters mix.

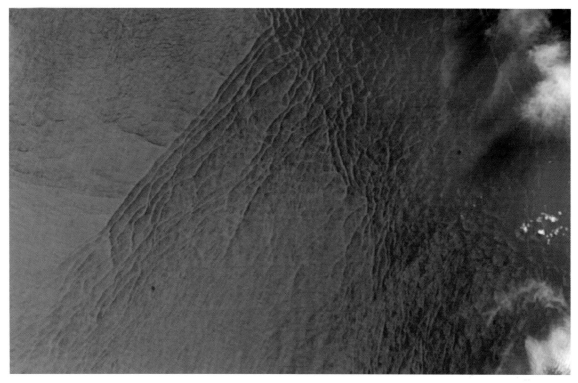

Pacific Ocean internal waves as seen from space. These waves cannot be seen with the naked eye and can stretch for miles and move through the ocean for several hours.

two layers of water is smaller than that between air and water, the restoring force for internal waves is smaller than that for surface waves. This means that it takes longer for water particles to return to their equilibrium level, so wave height, period, and wavelength can be larger. Internal wave heights are typically 50 to 100 times larger than surface wave heights. They may stretch out for miles and travel greater distances than most surface waves.

Although internal waves themselves are rarely visible from the surface, they can provide signs of their presence. As the crest of an internal wave nears the surface, it creates a slick, an area of relatively smooth water. Over the trough, water is rougher. These bands of rough or smooth water may stretch out for 60 miles (100 km). Planktonic animals are often concentrated in front of the wave, which draws in fish, birds, and mammals as well.

DEAD WATER

Sailors have long been aware of a phenomenon they call dead water. When traveling through a fjord or estuary, or near an ice shelf, a ship may suddenly slow or even stop for no apparent reason. It turns out that if the propellers of a ship reach the pycnocline level, the ship generates internal waves. Once this happens, energy from the propeller goes into generating waves and not into moving the ship.

Another mystery that may involve internal waves is the 1963 loss of the nuclear-powered attack submarine the USS *Thresher*. While traveling rapidly underwater east of Cape Cod, the *Thresher* reported "minor" distress. Shortly afterward, the submarine disappeared. The entire crew of 129 men was lost. Some experts attributed the accident to a large storm that had passed through the area two days before. The thinking was that the *Thresher* had been caught in undersea waves or eddies that carried it below its test depth before its crew had a chance to respond.

Tides, Sun, and Moon

Around the world, the sea rises and falls along the coast in daily cycles known as tides. Tides are caused by the gravitational pull of the Sun and Moon and modified by the geometry of continents and ocean basins. In addition to the up-and-down motion of water along the shore, tides can create stunningly rapid currents that switch direction with the tidal cycle. A rising tide creates flood currents, while a falling tide creates ebb currents. Slack tide, the time when currents are weakest, happens when the tide turns, or switches from ebb to flood.

THE PULL OF THE MOON

To understand what causes tides, it is easiest to start with a simplified case: the equilibrium theory of tides, which is based on a uniform Earth covered in one big ocean. This theory puts forth the following, simplified explanation of tidal movement: The gravitational pull of the Moon attracts both Earth and the ocean. Without a force acting against gravity, the Moon and Earth would crash into each other. What keeps them apart is inertia, the tendency of moving objects to travel in a straight line. On the side of Earth facing the Moon, the balance between gravity and inertia swings slightly in favor of gravity, creating a bulge of water toward the Moon.

Above: Industrial cranes being transported beneath the Oakland Bridge in San Francisco Bay. Careful planning and detailed knowledge of the tidal cycle allowed the cranes to clear the bottom of the bridge by a mere 6 feet (1.8 meters). Top left: Low tide on the Portsmouth, New Hampshire, seacoast.

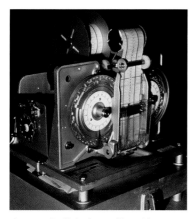

Automatic digital recording tide gauges like this one were used in the 1920s. An aluminum-backed paper tape was punched every six minutes to indicate tidal stages.

The Moon shines on Hobart Bay in Alaska. The gravitational forces of the Sun and Moon cause waters of the ocean to swell and recede at different parts of the Earth at different times of day.

LONG DAYS, DISTANT MOON

Although they might seem independent, the tides, the length of a day, and the distance between Earth and the Moon are all related to tidal friction. As the crest of the tidal bulge moves west, Earth spins east underneath it. This pulls the wave eastward and prevents the bulge from remaining directly opposite the Moon. The added mass of water in the bulge attracts the Moon, which speeds it up in its orbit, moving the Moon 3 inches (1.2 cm) farther from Earth each year. The Moon also pulls back on the bulge and slows the rotation of Earth by one to three milliseconds per century. Earth's drag on the Moon creates a phenomenon known as tidal locking, which means that only one side of the Moon ever faces Earth.

On the side of Earth opposite the Moon, the balance shifts in favor of inertia, creating a bulge of water away from the Moon. Earth completes one full revolution a day, so the theory continues that there should be two equal high and low tides each day. Because the Moon passes overhead an hour later each day, tides occur an hour later each day as well. Although we do not notice them, there are tidal bulges in Earth's crust as well as in the oceans.

THE EFFECT OF THE SUN

For tide-generating forces, proximity matters more than size. While the Sun's gravitational pull on Earth is 177 times that of the Moon, its tidal force is just half that of the Moon. When Sun and Moon are in line with each other, tides are more extreme because the gravitational pull of Sun and Moon can act in concert. These are known as spring tides; they occur twice a month, when the Moon is full or new. When the Sun and the Moon are at right angles to each other, the tides are at their weakest, because the gravitational pulls of the Sun and the Moon cancel each other out. These are known as neap tides; they also occur twice a month.

HOW HIGH?

Over the years, people have measured the height of the tides in various ways. Early devices had a pen-and-ink strip attached to a float inside a tube attached to a dock. Later, a punch strip that could be fed into a computer replaced the pen and ink.

Both of these designs required technicians to visit the tidal gauges regularly to make sure the devices were still working properly and to collect the data.

Now data are stored on microchips that can be fed directly to a computer, and satellites transmit data hourly. Rather than mechanical floats, modern tide gauges use acoustics, just as ships use acoustics to measure the depth of the sea. A machine at the top of a narrow tube emits a sharp ping, and then measures the return time. Tidal height measurements are generally taken every six minutes.

Tides in the Real World

The equilibrium theory of tides was fairly well settled in 1687 when Newton published *The Mathematical Principles of Natural Philosophy*, in which he discussed tide-generating forces. It was clear that something was missing from this explanation, however. Tides on Earth are higher than predicted and do not always follow the expected pattern. Some areas have just one high and one low tide a day, a pattern known as diurnal tides. Others have the expected two high and two low tides a day, but they are not always equal. If the pairs of high and low tides are roughly equal, tides are said to be semidiurnal. If the pairs are unequal—that is, if one high is higher than the other and one low is higher than the other—the tide is called mixed semidiurnal. Pierre-Simon Laplace, a French mathematician, took up this problem in 1775, developing the dynamic theory of tides. His theory has been expanded and refined over the years, improving the ability of scientists to predict tidal patterns around the globe.

SHALLOW-WATER WAVES

The key to understanding tides is the fact that tides are just very big waves. The combination of their length and the water depth causes tides to behave as shallow-water waves. These waves result when the water depth is less than one-twentieth of the wavelength, and the ocean is always less than one-twentieth of the tide wave's wavelength. Like tsunamis, tides are always affected by interaction with the ocean bottom. Because the tide wave is so big—just two crests on the entire Earth—the Coriolis effect also influences its movement. Its progress is also interrupted by continents, and the tide wave can refract and diffract just like any other wave.

In many ocean basins, the tides travel like a progressive wave, and it is possible to draw lines across the ocean indicating where the crest, or high tide, will be at set time intervals. In other cases, the tide wave forms rotary standing waves, that is, standing waves that rotate around a single, central node called the amphidromic point. Because the crests and the troughs of standing waves cancel each other out at nodes, amphidromic points have no tides. There are twelve or so of these points around the world, and large bays, such as

Above: The Gulf of Saint Lawrence is where the Great Lakes feed into the Atlantic Ocean via the Saint Lawrence River. There are two places in the gulf that have no tides due to the phenomenon of amphidromic points, which cause standing waves to cancel each other out. Top left: The Portland Head Light in Maine looks out over the Atlantic Ocean. Portland and most other points along the Atlantic coast experience semidiurnal tides.

A tidal bore in the Bay of Fundy, which has the world's greatest tidal range.

The San Francisco tide gauge house celebrated 150 years of continuous observations in 2004. Today, tidal predictions for years into the future are available on the Internet for locations around the globe.

WHAT IS NORMAL?

Tidal height is measured against some locally or regionally specified reference level generated by measuring tidal heights over a number of years. The reference level is different for different sites. Most tidal stations calculate the mean tidal level (the average of low and high tides) and the mean high- and low-tide levels. At locations with two different highs and lows a day, the mean higher high, lower high, higher low, and lower low are all calculated. Navigational charts generally list the average low or lower low-water level. So-called minus tides, when the tidal height falls below the mean low or lower low-water level, make for trickier boating but are great fun for marine biologists and beachcombers.

the Gulf of Saint Lawrence, may have their own amphidromic points. The Coriolis effect causes the tides to rotate clockwise around amphidromic points in the Southern Hemisphere, and counterclockwise in the Northern Hemisphere.

TIDAL BORES

In narrow bays, the height of a tide wave may be amplified by the way the wave bounces around within the basin. The confining bay limits reflection of the tide wave to the head of the bay, which can make the difference between high and low tides, the tidal exchange, much greater there than at the entrance. Canada's Bay of Fundy, for instance, has a tidal exchange of around 6.5 feet (2 m) at its mouth and a range of 35 feet (10.7 m) at its head. In areas with a large tidal exchange and a shallow bay or river mouth, the incoming tide may create a tidal bore. The shallow conditions cause the tidal crest to steepen, while the confinement of the bay or river mouth makes it move at a speed greater than a shallow-water wave normally would at that depth. The crest breaks to form a spilling wave front that moves up the bay or river at speeds of up to 25 miles per hour (40 km/h). While tidal bores are usually 3 feet (1 m) tall, they may reach heights of 26 feet (8 m). Tidal bores occur regularly in the Bay of Fundy, southwestern China, the Amazon, the Ganges Delta, and the Severn River in the United Kingdom.

Getting Electricity from the Tides

As the negative effects of fossil fuels become clearer and supplies dwindle, people are becoming more interested in alternative sources of energy. Each region of the globe has renewable resources it can bring into play: sunlight, wind, geothermal energy, and, in some locations, tidal power.

Traditional tidal power plants involve building a dam across a narrow tidal inlet. Just like dams on rivers, they allow water to build up on one side, then use the potential energy generated by the difference in water height on different sides of the dam to run turbines. Simpler systems have turbines that spin in only one direction. Water is allowed to flow freely through the dam on the flood tide and runs the turbine on the ebb. Other systems have two-way turbines that generate power from both ebb and flood tides.

HOW MUCH POWER?

The oldest and largest tidal power plant in the world was built in 1966 in France's La Rance estuary. With a dam 2,800 feet (850 m) long holding 24 turbines generating power on both ebb and flood tides, the plant has a capacity of 240 megawatts. The Annapolis River plant in Nova Scotia, the largest straight-flow plant, has a capacity of just 20 megawatts. China has eight tidal power plants, with a total capacity of 6.12 megawatts, although not all plants are still in use. Smaller pilot plants have been built in a few other countries, and several countries have plans for plants with capacities of greater than 7,000 megawatts.

Despite these promising numbers, there are several obstacles to turning tidal power into a major energy source. To build tidal power plants, the tidal exchange must be great enough and the channel narrow enough to generate a sufficient current. Further, these conditions must be met in an area

Above: The world's first commercial-scale wave power station resides on the Scottish Hebridean Island of Islay. This station generates 500 kilowatts of electricity, which is enough to power 300 homes. Top left: Industrial pollution is one result of reliance on fossil fuels that has led to increased interest in alternative energy sources.

An illustration showing marine current turbines at work. These turbines act like submerged windmills that are driven by flowing water rather than air. Using tidal power rather than wind or wave energy makes this technology less susceptible to the quirks of weather. The first turbine to generate power has been installed in Devon, England.

close enough to population centers to make transporting electricity between the plant and the users worthwhile. To date, fewer than 300 possible sites for such plants have been identified worldwide. Still, some countries could meet a significant portion of their energy needs with tidal power. The United Kingdom, for instance, has eight possible sites that together could supply 20 percent of its energy needs.

WATER MILLS

An alternative to the dam-and-turbine strategy is marine current turbines, essentially underwater windmills that use tidal currents or other strong ocean currents to generate power without need of a dam. Because water is much denser than air, a turbine with the same size blades would generate the same amount of power in a 2-knot current as it would in a 20-knot wind. Prototype 300-kilowatt turbines were installed off the coast of Devon, England, and Hammerfest, Norway, in 2003, and several companies have plans for building full-scale "tide farms."

THE DOWNSIDE

In addition to practical concerns about tidal energy, some people have also raised environmental concerns. Tidal dams make it harder for many animals to migrate in and out with the tide, although as with hydroelectric dams, special passageways for animals could be built into the dam. By essentially reducing the size of the bays in which they are built, dams can increase the tidal range and current speeds, causing landowners near proposed tidal power plants to worry about increased erosion. Because the rotors of marine current turbines turn at about a tenth the speed of boat propellers, it is unlikely that they pose a significant hazard for underwater animals.

CHAPTER 8

AT THE OCEAN'S EDGE

Left: Throughout history, people have lived at the ocean's edge, both for practical and aesthetic reasons. Today many people pay exorbitant prices for an ocean view. Top: Coasts are formed from many different materials. On the island of Hawaii, lava has flowed all the way to water's edge in some places, creating a new landscape. Bottom: The Caribbean island of St. Lucia depends heavily on its coastal and marine resources for the success of its tourism industry. As is often the case in such places, the overuse of these resources has compromised them.

Coasts are dynamic places. Wind, waves, rain, and rivers constantly reshape them, taking away material here, adding it there. The shape of a coastline depends on how it was originally formed and the material it is made of, such as the hard rock of a lava flow or the soft mud and sand of a river delta. The shape and character of a coastline also bear the mark of what has occurred since it was formed, a story that can go back thousands of years. Coastlines are also shaped by the rise and fall of sea level, which has gone up and down by hundreds of feet over the millennia.

Coasts are also where most people experience the sea. People live, walk, swim, and gather food there. Human activities shape and are shaped by the coast. Coasts are the starting and ending points of many journeys. In the words of ecologist Rachel Carson, "To stand at the edge of the sea . . . is to have knowledge of things that are nearly as eternal as any earthly life can be."

Whether a sandy or rocky beach, a saltwater marsh or coral reef, the ocean's edge is a diverse and fascinating ecosystem. It is a place where the forces of land and the forces of the ocean meet.

Before the Waves

Some of the most visually stunning coasts in the world owe their dramatic character not to waves and tides, but to nonmarine processes. The United States's Tomales Bay and Mexico's Gulf of California lie along the San Andreas Fault, created by the sliding of one tectonic plate against another. Sand dunes along these shores are shaped by wind. In volcanically active regions such as Hawaii and Iceland, fresh lava-flows regularly add new coastline, and eroded lava can create distinctive black sand beaches. If a volcanic island explodes and ends up below sea level, or loses one of its sides, the result is a beautiful, curving bay. These coastlines, shaped more by terrestrial than marine processes, are called primary coasts.

GIVE AND TAKE OF RIVERS

Two hundred million years ago, Sydney Harbor in Australia was the mouth of what scientists have named the Parramatta River. The river deposited tons of sediment that became the thick layers of sandstone currently visible in the headlands that guard the mouth of Sydney Harbor. Tectonic forces lifted the sandstone up, and the river eroded channels into the soft rock. At the end of the last ice age, sea level rose and flooded the area with seawater. The drowned river valleys now form the complex system of smaller bays and harbors that contribute to Sydney's charm. Chesapeake and Delaware bays in the United States are also drowned river valleys, which can be identified by their characteristic patterns of branching V-shaped channels.

Rivers create as well as erode land: An average of 530 tons per

Above: Black sand beaches, which are made of eroded lava, are found throughout the Hawaiian Islands. Above right: Sandstone cliffs at the entrance of Australia's Sydney Harbor are the remains of sediment deposited there 200 million years ago. Top left: Sand dunes are coastal features that are shaped by wind rather than water.

second of sediment is carried to the sea by rivers, leading to the formation of extensive deltas and alluvial plains. These regions are generally flat and biologically rich, such as the Mississippi Delta in the United States or the Nile Delta in Egypt. The sediment from deltas also contributes to the formation of beaches, sandbars, and barrier islands.

FORMED BY ICE

When glaciers cut through hard, rocky, mountains, they create distinctive U-shaped valleys. When the bottom of these valleys is below sea level, the result is the narrow, deep, and steep-sided fjords characteristic of many high-latitude coastlines around the world. Where glaciers push into the ocean, these valleys may form below sea level. In this case, the ocean floods the valley as soon as the glacier retreats. Alternatively, valleys may also form above sea level and become flooded only as glaciers melt and sea level rises in times of warming climate.

Glaciers also create terminal or lateral moraines, collections of rock, cobble, and sand that are ground down or carried by glaciers and deposited at their edges or ends. The result is a sandy, porous region somewhat higher than its surroundings. In the United States, both Long Island and Cape Cod are terminal moraines left behind after the last ice age.

Glaciers also influence coastlines indirectly through their local and global effects on sea level. During ice ages, more land is exposed to the terrestrial forces of wind and water, creating features that may remain when sea level rises as the ice age ends.

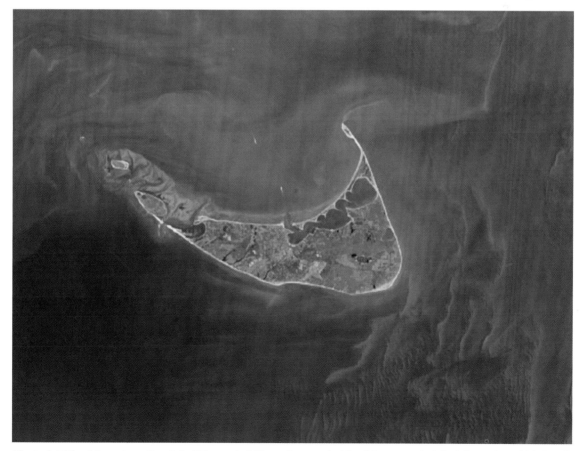

Nantucket Island from space. Located off the coast of Massachusetts, the island is composed of glacial moraine, which is rock, cobble, and sand that was left behind by melting glaciers in the last ice age.

The Anatomy of a Beach

The shape, slope, and composition of a beach can tell stories about the beach's history, and about what organisms are likely to live there. They can also help predict the future of the beach, important information in considering whether or not to build along particular stretches of shoreline. Beaches are often categorized by what they are made of, which can be anything from shells and coral to quartz or lava. They are also categorized by how big the material that makes up the beach is, be it boulders, cobbles, sand, or mud. As a rule, fine sandy beaches are broad and gently sloping, while cobble beaches tend to be steeper.

BEACH TOPOGRAPHY

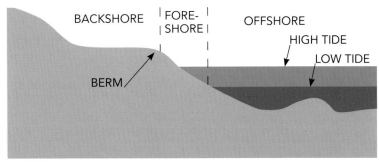

This diagram illustrates the basic parts of a beach. The boundary between backshore and foreshore is marked by a berm. Foreshore and offshore limits are determined by where on the shore low tide reaches.

NAMING THE PARTS

The three basic regions of the beach are the backshore, covered only during the highest tides, the foreshore, which is covered and uncovered daily by the tides, and the offshore, which is below the level of low tide. There may be one or more terracelike structures known as berms near the

Below: A shoreline exhibits gradation of sediment from coarse pebbles at water's edge to sand. The deposit of seaweed along the beach shows the level of the last high tide. Top left: Mount Desert Island in Maine.

Rip current warnings should not be taken lightly. These are powerful currents of water flowing away from shore that typically extend from the shoreline, through the surf zone, and past the line of breaking waves. Rip currents can occur at any beach with breaking waves, including the Great Lakes. They account for over 80 percent of annual rescues performed by surf beach lifeguards.

boundary of the foreshore and backshore. Berms run parallel to the shoreline and are built by the deposition of material by the highest tides. In summer, beaches may have two berms. The higher one is the winter berm, created when winter storms pushed the waves higher up the beach. The lower, or summer, berm usually disappears in winter when the powerful waves either push the material up to the winter berm or carry it out to sea. Typically, on the seaward side of each berm is a steeply sloping area known as the beach scarp.

Just below the average low-tide level there may be a flat area known as the low-tide terrace. Between this and the rest of the beach is often a steep area known as the beach face. Further offshore, water movement creates bars and troughs that move seaward in winter and shoreward in the summer months.

CUSPS AND RIP CURRENTS

Some beaches have distinctive cusps along their length, a scalloped pattern in which the beach alternately juts out farther or less far into the water. The mechanism for the formation of cusps is still hotly debated. Some evidence suggests that cusps result from the formation of a particular pattern of standing waves, while other evidence points toward the interaction of local morphology, currents, and sediment.

Waves push water up onto the beach, and the water has to get back to the sea somehow. Depending on a variety of factors, it may flow out relatively evenly across the beach, or it may form strong, localized offshore flows called rip currents. Being caught in a rip current can be terrifying, but the solution is simple: Swim parallel to shore rather than straight back to shore against the rip current. Swimming against the rip current is exhausting, while swimming parallel to shore moves a swimmer out of the rip current into an area where getting back to shore is easier. Also, onshore eddies often form along the edge of rip currents, which can help ease a swimmer to shore.

Shifting Sands

Coastal erosion is a fact of life. The daily pounding of the surf eats away at the shore little by little, and powerful storms can make whole islands disappear overnight. Wave action slowly straightens shorelines that are made of relatively uniform material: Energy is concentrated on areas of land that stick out into the ocean called headlands, causing them to erode more quickly than nearby bays. If the shoreline has areas of harder and softer material, however, those differences become increasingly visible in the contours of the coast. Hard, rocky areas erode more slowly, forming headlands, while areas of softer rock or sand erode more quickly, creating sandy coves or pocket beaches. If a rock formation is surrounded on all sides by softer material, erosion creates sea stacks, dramatic pinnacles of rock sticking up from the beach. Uneven rates of erosion are also responsible for the formation of sea caves and arches.

BARS AND BARRIERS

What happens to material that is eroded away from the shoreline? It may be carried offshore and deposited in long sandbars running parallel to the coast. Depending on the volume of material deposited, which can change with the seasons, sandbars may become big enough to break the surface, creating barrier islands. If barrier islands remain above the waves for long enough, characteristic ecosystems develop, including grassy dunes, back-barrier marshes, and maritime forests. Although vegetation can stabilize the island to a certain extent, barrier islands remain dynamic regions, reflecting the constant flux of erosion and deposition, waves and currents. Barrier islands act as natural breakwaters for the coast behind them, but as Hurricane Katrina proved, even large barrier islands can disappear in a matter of hours.

Depending on the pattern of waves and currents, sand spits and hooks rather than islands may form. In this case, deposits of eroded material remain connected to the mainland, forming natural harbors. If a spit grows long enough, it may completely enclose an area of water formerly open to the sea, creating a lagoon. Water in the lagoon may still rise and fall with the tide as water percolates through the loose material of the spit. Sea stacks and islands are sometimes attached to the shore by spits. These connecting landforms, known as tombolos, form when the refraction of waves by islands

Above: Sea stacks are common sights along the northwest coast of Oregon. Top left: Canaveral National Seashore Park is located on a barrier island along Florida's central east coast that separates the Atlantic Ocean from Mosquito Lagoon.

The force of the waves has carved this sea cave out of lava rock.

and sea stacks creates a low-current area between them and the shore.

ERODING ARCTIC

In the Arctic, coasts have generally been protected from the erosive force of winter storms by sea ice. Rather than breaking on shore, waves expend their energy at the edge of the sea ice, which can extend for miles from the shore. Because of global warming, the extent of sea ice is shrinking, and the ice breaks up and moves offshore much earlier than it had previously. Loss of protective sea ice, combined with thawing permafrost and rising sea level, has resulted in a dramatic

Ediz Hook from the air. The sheltered harbor on the right could disappear along with Ediz Hook if nothing is done to save it.

OF DAMS AND HOOKS

Washington State's Olympic Peninsula makes up the most northwesterly part of the mainland United States. Famed for its temperate rain forests, the peninsula is crisscrossed with streams and rivers, including the mighty Elwha. For hundreds of thousands of years, the Elwha carried sediment to the sea, creating a hook-shaped spit known as Ediz Hook. The Elwha was dammed to provide power and water in 1911, and much of the sediment that used to feed Ediz Hook was now trapped behind the dam. Sediment flow to the spit was further reduced as people built bulkheads along the shoreline, cutting off flow of material from eroding cliffs. Ediz Hook, home to a paper mill, a Coast Guard station, and numerous other businesses, began to disappear. In the 1950s, large boulders were placed along the seaward side of the spit to slow erosion, and since that time the annual cost of preventing the slip from disappearing has grown to around $30 million. The Elwha's dams are scheduled to be removed beginning in 2008 to help failing salmon populations, but dam removal may help Ediz Hook as well.

increase in erosion rates along much of the Arctic coastline. Near the village of Shishmaref, Alaska, land is disappearing at a rate of 10 feet (3.3 m) a year. In just a few decades, the village has gone from being hundreds of feet away from the water to being right on the edge of the sea. Every year, villagers are forced to move houses from the seaward side of the village to avoid losing them to the waves. Villagers would like to stay in their ancestral village, which archaeologists think has been inhabited for 4,000 years, but this may no longer be possible.

Cycling Sea Levels

While scientists debate the exact timing and details, all agree that sea level has changed significantly over time. In the past two million years, sea level may have been as much as 65 feet (20 m) higher than it is today, and as much as 410 feet (125 m) lower. As the last ice age waned, sea level rose an average of half an inch (1 cm) per year. The current rate of sea level rise is roughly 0.11 inches (3 mm) a year, slower than at the end of the last ice age but faster than most of the past century. By 2100, sea level will be from 1 to 3.5 feet (.30–1 m) higher than it currently measures.

GLOBAL CHANGES

During ice ages, more water remains on land as ice, reducing both the amount of water in the ocean and the global sea level. Climate warming melts the ice and adds to the volume of the ocean. If the East Antarctic ice sheet, the largest in the world, melts, it could raise sea level by more than 200 feet (60 m). Melting of the Greenland or West Antarctica ice sheets could raise sea level by 21 feet (6.5 m) and 26 feet (8 m) respectively. Ocean volume also changes due to water's expansion and contraction as it warms and

Above: While some coastal cities, such as New Orleans, are facing higher-than-average rates of sea level rise, others, such as Lysekil, Sweden, (shown here), have lower-than-average rates due to isostatic rebound following the melting of glaciers. Top: An iceberg near the glacier, in Greenland. The Ilulissat glacier, which has been studied by scientists for 250 years, has shrunk significantly in recent years, which is seen as evidence of global warming. Top left: Riggs Glacier in Glacier Bay, Alaska.

cools. A 1.8°F (1°C) increase in the average ocean temperature would raise sea level by about 24 inches (60 cm).

Global sea level can also change if the size of the ocean basin itself changes. This is easiest to imagine first at a regional level: Significant sediment accumulation in a relatively enclosed basin may cause sea level to rise, just as adding a large amount of sand to a swimming pool would raise the water level. On a global level, periods of rapid seafloor spreading lead to shallower ocean basins, causing sea level to rise.

LOCAL AND REGIONAL CHANGES

The growing and shrinking of glaciers has local as well as global effects. Because continents are essentially floating on the fluid mantle below, they rise and sink as weight is added or taken away, just as a boat rides higher or lower as cargo is loaded and unloaded. If land is covered with glaciers, the added weight causes the land to sink, leading to local sea level rise. When the ice melts, the land rises, causing a drop in local sea level. This process, which is called isostatic rebound, can take many thousands of years. For example, as the last ice age ended about 10,000 years ago, parts of northern Europe and North America rose at rates of approximately 3 inches (7.5 cm) per year. Average uplift rates have since slowed to half an inch (1 cm) or less a year. There are still locations, however, that continue to rise fairly dramatically. Scandinavia, for example, still rises at a rate of 0.4 to 2 inches (1–5 cm) per year.

Erosion and accumulation of sediment can have similar effects. As sediment builds up, land sinks under its weight. Sea level can remain stable if sediment input keeps pace with sinking, but human activities often shift this balance. The Mississippi River Delta had high levels of past sediment input, and is continuing to sink in response. Because the sinking of continents into the mantle is slow, reaching a new equilibrium level for the delta will take centuries. The construction of dams, levees, and other structures has decreased sediment input significantly, however, so the surface of the land is no longer being built up, and local sea level is rising. The withdrawal of groundwater for agriculture has further hastened the sinking of the land. Local sea-level rise in the delta is consequently rapid: 40 or 50 square miles (100 to 130 square km) of land are lost each year.

Tectonic forces such as earthquakes or the collision or separation of two continents also influence regional sea level by moving land up or down. The level of the sea surface in a particular region can be affected by changes in atmospheric pressure such as those typical of El Niños. For instance, if atmospheric pressure rises in a region of normally low pressure, sea level will drop.

Erosion of a cliff reveals buried piles of oyster shells.

LOOKING INTO THE PAST

Scientists study past sea levels using many lines of evidence. Along many coastlines, old shoreline features such as wave-cut terraces and beach deposits are now high and dry. By dating these deposits, scientists can tell when sea levels were that much higher than they are today. Scientists can also trace rising and falling sea levels by looking at the sequence of sediment or rock types. A period of rising seas, for instance, may leave a layer of marsh deposits covered by sand deposits covered by deeper water deposits, while falling sea level would leave a reverse sequence. Microscopic fossils in the sediments below the Greenland and Antarctic ice sheets also provide evidence for when the region was covered with seawater and when it was high and dry. In warmer regions, living and fossilized shallow-water corals provide good indicators of sea levels past and present, since these animals have a fairly restricted depth range.

Living Rocks

Shallow-water coral reefs create living coastlines. They are popular diving and snorkeling destinations, generating billions of dollars in tourist revenue yearly. Reefs support thriving fishing industries and provide food and income to local communities. Reefs also help to protect the shoreline and communities from storms and flooding, preventing significant amounts of erosion and property loss.

The reef structure is a calcium carbonate skeleton created by coral, small polyplike animals resembling sea anemones. Some coral species live as individual polyps, while others form interconnected colonies of genetically identical individuals. Corals use stinging tentacles to capture food and to fight with other corals over space. The often brilliant colors of the reef come from photosynthetic organisms called zooxanthellae that live inside the cells of the coral animals. If they lose these tiny algae, the corals turn white in a frequently fatal process known as bleaching. Coral reefs are home to 25 percent of known marine species.

TYPES OF REEF

There are three basic types of reef: fringing, barrier, and atoll. Fringing reefs form around the edge of land and connect to the shore. They thrive only if rainfall and runoff are low, so are often found on the leeward side of islands. Barrier reefs are separated from the shore by a lagoon. The lagoon may be small and shallow, or up to 200 feet (60 m) deep and 190 miles (300 km)

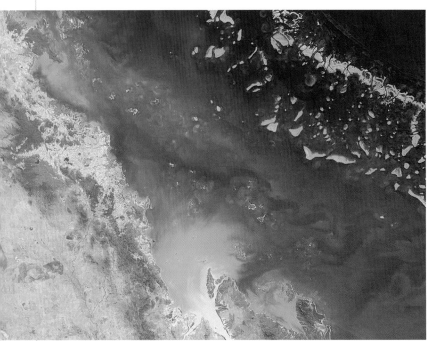

Above: Approximately 360 species of hard corals can be found in Australia's Great Barrier Reef. The oldest individual coral on the reef is believed to be 1,000 years old and is a member of the genus Porites, *commonly known as boulder coral. Top left: The sand-colored area around this island is a fringing reef. These reefs develop in shallow waters a short distance from shore.*

A bleached coral. High temperatures, disease, and other sources of stress can lead corals to expel the algae that are responsible for a coral's bright colors.

wide. Australia's Great Barrier Reef, the largest animal-made structure on Earth, is composed of thousands of reef segments with a total area of 135,136 square miles (350,000 sq km). It is younger and thinner at the southern end, perhaps because the tectonic plate on which the reef sits is moving north toward the equator, constantly bringing new stretches of ocean into conditions warm and sunny enough to support shallow-water reefs. Atolls are the classic ring-shaped coral islands surrounding a partially or completely enclosed lagoon. The biggest is the Kwajalein Atoll in the Marshall Islands. More than 176 miles (280 km) around, it surrounds a 1,100-square-mile (2,850 sq km) lagoon. Darwin suggested that atolls are former fringing reefs around now sunken islands, a plausible explanation given our current understanding of island formation and plate tectonics.

THREATS TO CORAL REEFS

Myriad threats face coral reefs today. Some, such as hurricanes, are natural threats that have been around as long as corals themselves. Others are new, the result of direct or indirect human action. By some estimates, 28 percent of coral reefs have died or are severely degraded.

Coastal construction can increase the outflow of freshwater and sediment, potentially smothering coral. Increased nutrients from agriculture, livestock, and sewage can fertilize algae that outcompete corals for space or light. And while sustain-

Scuba divers can help protect coral reefs by avoiding direct contact with the organisms that live at diving sites.

WHAT YOU CAN DO

Individuals can do a lot to help protect coral reefs. As consumers, they can make sure that any reef animals or coral products they buy have been harvested in a sustainable fashion. This is true whether the animals are destined for an aquarium, the dinner table, or a display case. People who go out in boats around reefs can ask boat operators not to anchor on the reef, and make sure that boats are well maintained to minimize leaking gas or oil. Divers and snorkelers can avoid touching the coral. Some corals are sensitive to chemicals found in sunscreens and lotions, and even small amounts of abrasion and contamination can have big effects in major tourist areas. People who live near reefs or rivers that flow toward reefs can minimize their use of pesticides and fertilizers, and visitors to these areas can seek out hotels that operate in a reef-friendly fashion.

able fishing is a feasible alternative, some people still choose to use cyanide or dynamite to collect fish for both the food and the aquarium trade. These practices kill coral and other animals as well as destroy the structure of the reef. Even tourists who are attracted by coral reefs may contribute to their decline. Many visitors collect or purchase pieces of coral as souvenirs, and tourist boats can damage corals with their anchors or by polluting the water.

A more wide-scale threat comes from climate change. While zooxanthellae need light to survive, too much light and heat can be deadly. If conditions become stressful, zooxanthellae disappear and the coral turns white. Sometimes corals can recover from this bleaching, but often they do not. Stress can also cause outbreaks of potentially fatal coral diseases. Scientists are trying to identify factors that can help corals avoid or recover from bleaching and disease.

Where Rivers Meet the Sea

Estuaries, enclosed or semi-enclosed places where rivers flow into salt water, take on a variety of forms, but all are rich and teeming with life. They may be deep fjords cut into rocky hills by glaciers, or shallow bays on coastal plains. They may form behind sandbars or in cracks or folds caused by tectonic activity. They may receive large and steady volumes of freshwater from a large river, or just small or seasonal inputs from smaller ones. The mixing of fresh and salt water, of nutrients and organic material from the land with marine ecosystems, creates a unique suite of physical, chemical, and biological processes. Estuaries provide habitat for three-quarters of the world's commercial fish catch, supporting millions of jobs and billions of dollars in the fishing industry. The marshes and wetlands that surround estuaries act as water filters and buffers against erosion and flooding. Estuaries are popular tourist destinations as well.

TYPES OF ESTUARIES

When freshwater flows into salt water, any of several things may happen. Because freshwater is less dense than salt water, the freshwater tends to float on top of the salt water. If the input of freshwater is large relative to the tidal exchange, such as at the mouth of a large river, a relatively fresh surface layer will flow out toward the sea, while a wedge of salt water creates a deeper landward current. The salt wedge moves in and out with the tide, and may travel quite far inland. In the Hudson River in New York, the tides can influence river flow as far up the river as 150 miles (241 km).

If freshwater input is relatively small compared with the tidal exchange, a well-mixed estuary forms. There is no difference in salinity with depth, although a salinity gradient forms along the length of the estuary. Usually, salinity is highest near the ocean, becoming progressively lower farther up the estuary. If freshwater input stops and evaporation is high, however, the water at the head of the estuary becomes saltier than the water at the mouth, creating what is called a reverse estuary. Well-mixed estuaries are typical of shallow bays.

In areas with an intermediate rate of freshwater input, a

Above: Pine Island is part of a National Estuarine Research Reserve outside of Charleston, South Carolina. The state has approximately 345,000 acres of salt marsh. Top left: Cordgrass is the dominant plant in this salt marsh. Marshes sometimes surround estuaries and provide protection against erosion and flooding.

partially mixed estuary forms. Some mixing occurs between the salt and fresh layers, but a noticeable gradient in salinity from surface to bottom remains.

Estuaries that form in glacier-cut fjords are narrow, steep-sided, and usually have a shallow sill at their mouth, which inhibits water exchange at deeper levels. Because of this, the boundary between fresh and salt water may be very sharp, and the deeper part of the estuary may become stagnant.

FLUSHING THE ESTUARY

Depending on freshwater inflow, tidal exchange, and a variety of other factors, water may be retained in estuaries for longer or shorter periods. The average duration that water stays in the estuary is known as the residence time. Different

parts of an estuary may have longer or shorter residence times. In Puget Sound, along the west coast of North America, residence times in isolated arms of the estuary are six to twelve months, compared to two months in the main basin. Regions with longer residence

times take longer to get rid of pollutants and are more prone to problems with low oxygen levels. Residence time changes seasonally depending on river flows: When river flow is high, residence time is reduced. Periods of drought or flooding also have strong effects.

Above: Estuaries are among the most biologically productive ecosystems on Earth. The Puget Sound in Washington State is home to any number of fish, mammal, and bird species. Salmon travel downstream to this estuary, where they feed and grow in preparation for their ocean journey. Top right: The estuarine habitat of the Tracy Arm Fjord in Alaska provides mountain goats, sea lions, and whales with seasonal respite. It is also a popular cruise ship destination.

Coastal Development

By some estimates, over half of the world's population lives on or near coasts. People live on coasts for practical and aesthetic reasons. Ports provide jobs, fishing provides food and employment, and the sound and smell of the sea can be soothing or inspiring. But shorelines are unstable places. Beaches are in a constant state of flux, and cliffs erode, undercutting housing and other structures. Humans have tried to tame natural processes over the years, with mixed success. A quick overview of sediment transport along coastlines will clarify why controlling erosion and waves is not so simple.

MOVING ALONG

The swash, or movement of water onto a beach, is stronger than the backwash. If waves hit a beach straight on, there will be net onshore transport of materials. If they hit the beach at an angle, the net movement of water and material will run parallel to the coast, a phenomenon called longshore transport. Shorelines have a characteristic longshore transport direction, determined by the direction of the waves typical for that shoreline. In the United States, the flow is usually north to south, following the prevailing wind direction.

The movement of sediment is broken up into distinct drift sectors. At one end is an area of net erosion, often characterized by eroding sea cliffs. Material from this spot is transported along the shore until it reaches a low-energy area such as a bay or harbor, where it gets deposited, marking the other end of the drift sector. In between is a dynamic equilibrium between deposition and removal of material from the beach. While beaches in this region may not appear to change much, large volumes of material are constantly moving along the shore.

Left: Houses built on stilts are particularly vulnerable to the effects of erosion. The erosive forces of a severe storm compromised this house. Top left: Fishing ports have sustained the economies of seaside villages around the world for centuries.

INTERRUPTING THE FLOW

Just how much material is being transported is easy to see when something blocks the flow of material through a drift sector. The construction of jetties and other shoreline structures has provided ample opportunity for such measurements.

Jetties are walls built perpendicular to the shoreline to protect harbor mouths or other features. Because they stick out from the shoreline, they block longshore transport. On the upcurrent side, sand builds up and enlarges the beach, while beaches are starved of sediment and disappear on the downcurrent side. Jetties in areas with average rates of sediment transport can trap around 30,000 dump-truck loads (240,000 cubic yards, or 183,493 cubic meters) of sand each year. Jetties in areas with high rates of sediment transport may trap hundreds of times that amount.

Breakwaters, which are built offshore and parallel to beaches, can also create sediment traps. Behind a breakwater, longshore transport is weakened, and material builds up, while sediment transport to downshore areas is reduced. Eventually, the beach will build out as far as the breakwater extends.

Not surprisingly, people down the shore from sediment-trapping structures try to stop their beach from disappearing. A common approach is to build groins, relatively short walls built perpendicular to the shoreline. While groins trap sediment on the upcurrent side, they accelerate loss of beach on the downcurrent side. As deposition decreases, erosion becomes the dominant force, threatening houses and other structures that had been relatively safe. People in these areas may build seawalls with the idea of protecting their property, but this can make things even worse. Seawalls block the transport of material from land to beach, speeding up the loss of beach in front of the seawall. This increases wave force on the seawall, hastening its demise and increasing the chance that waves will wash over the top of the seawall, eroding the material behind it. This can lead to the collapse of seawalls, which are not built to stand unsupported from both sides.

Rock seawalls are often built to slow the process of erosion. Unfortunately, this approach is not always effective and can sometimes even speed up the process.

WORKING WITH NATURE

Some governments and private groups have decided to work with rather than against natural coastal processes. In the United States, Rhode Island banned the construction of seawalls, jetties, and similar structures along much of its coastline. Residents of Santa Barbara, California, purchased 69 acres of undeveloped shorefront property to serve as an erosion buffer zone, protecting the homes behind it. Members of another California community decided to remove rather than rebuild a bike path and parking lot damaged by erosion, instead restoring the natural sand and cobble buffer zone. In the United Kingdom, the British Environmental agency chose to "[help] nature protect the coastlines" by removing some existing shoreline structures and setting seawalls farther back from the shore.

A view of Long Island Sound from Block Island, Rhode Island. Residents here have taken action to protect the coasts of the Ocean State.

CHAPTER 9

LIFE BETWEEN THE TIDES

Left: Mudflats form where wave energy is low, and host a variety of burrowing shellfish and worms. Top: In areas with powerful waves, cobble beaches make unstable habitats and may have little life. Bottom: While this fish needs to stay underwater, its anemone host may be exposed to air for hours each day.

Around the globe, tides rise and fall with the cycles of the Moon and Sun. As they do, they alternately flood and expose the land at the ocean's edge, an area called the intertidal zone. In some regions the difference in water level between high and low tide is so small that the intertidal zone is almost nonexistent. In others, it can extend over 30 feet (almost 10 m), generating impressive tidal currents and an ever-changing mosaic of water and land, algae and animals. In such areas, the shore displays a remarkable transition zone between the terrestrial and the marine. Its inhabitants must be able to handle a life of extreme fluctuations, alternating between the open air and seawater submersion once or twice a day. Yet handle it they do, creating a vibrant ecosystem in the process. What that ecosystem looks like varies tremendously with climate, the extent and timing of tidal fluctuations, the presence or absence of rivers, and as a function of how hard or soft, muddy or sandy the bottom is. From tropical reefs and mangrove forests to temperate salt marshes and rocky beaches, there is no better place to begin an exploration of marine life.

Organisms and Interactions

The diversity of life, by many measures, is greater in the ocean than on land. Although there are more named terrestrial than marine species, this probably reflects differences in time spent exploring the two realms rather than in the actual number of species. Of the forty or so animal phyla, only one is strictly a land-dweller (the Onycophora, or velvet worms), while a dozen or so are found only in the sea. Thus, a trip to the intertidal zone can open one's eyes to just how many ways animals can live, feed, and reproduce.

THE CAST OF CHARACTERS

Most marine primary producers (organisms that harvest energy from the Sun through photosynthesis) are algae rather than true plants, meaning they lack any specialized tissues for transporting fluid and nutrients. Some are big, such as 150-foot-tall giant bull kelp, while others are microscopic. While trees can live for over a thousand years, even the largest algae often live for just a year or two. Single-celled algae have even shorter lives, reproducing by splitting in two as often as once a day.

Intertidal animals are likewise varied. Some predators, notably many crabs and fish, hunt their prey visually, while

Above: Seaweeds are amazingly diverse. They create forests above, a grassy turf below, and everything in between, providing food and shelter for numerous other organisms. Top left: A porcelain crab makes its home amid the stinging tentacles of an anemone.

others use chemical cues. Most digest food internally, but a few, like sea stars, turn their stomach inside out and digest their prey externally. Bottom-dwelling herbivores, like limpets, crawl, eyeless, in search of algae. Others, including microscopic shrimp relatives called copepods, swim about capturing floating single-celled algae. Because seawater, unlike air, is full of potential food items, many inter-tidal animals take an approach to eating that is not possible on land: suspension feeding. Eschewing pursuit, suspension feeders stay put and pull food from the surrounding water. Some, such as oysters, suck in gallons of water a day, sorting out particles internally. Others, such as barnacles, stretch feathery appendages out into the water to catch and transport individual food items.

Let us not forget bacteria. They provide a rich food source for other microscopic life and help to decay organic matter and recycle nutrients. There are millions of times as many bacteria in the ocean as there are stars in the known universe.

This anemone gets its bright green color from tiny algae that live inside its cells. Anemones kept in the dark will lose their algae and turn white.

INTERACTIONS

Perhaps because of all this diversity, species in marine food webs tend to be connected to more other species than is typical in terrestrial food webs, or more simply, organisms eat and are eaten by a greater variety of spe-cies. Marine food webs also have more omnivores—organisms that eat both animal flesh and algae or plants—and more cannibals.

Eating and being eaten are just two ways organisms interact. Individuals compete fiercely for space, food, and other resources. They can also make life easier for each other. For instance, algae can provide a variety of animals with shade and protec-tion from predators. When the interaction between two organisms of different species is particularly intimate, scientists call it a symbiosis. Parasitism is a symbiosis in which one partner benefits at the expense of the other, such as when a tapeworm lodges in a fish's gut and diverts nutrients from the fish for its own growth. Mutualism is a symbiosis in which both part-ners benefit, as when coral get nutrients from microscopic algae inside their cells and shelter the algae from herbivores and excessive light. A third type of symbiosis is commensalism, in which one partner benefits and there is little or no effect on the other. A number of small ani-mals, for instance, live inside marine sponges, using them for shelter with no apparent effect on the sponge itself.

Oysters are very effective at filtering algae and other small particles out of the water for food, and dense oyster beds, like this one in Washington's Willapa Bay, can play a key role in keeping estuaries clean and clear.

To Each His Zone

Astriking feature of the intertidal zone is how its inhabitants change over short distances, forming distinct bands that parallel the shoreline. This so-called zonation is most apparent in rocky areas, but is found in other habitats as well. Even apparently featureless sandy beaches can have marked shifts from one set of organisms to another along the gradient from high to low tide lines, as a careful sifting of the sediment reveals. What restricts creatures to their particular zones, keeping them from living higher or lower than they do?

A TALE OF TWO BARNACLES

Ecologist Joseph Connell did some elegant experiments to address this question. A common sight along rocky shorelines in the Northern Hemisphere is a band of little brown barnacles called *Chthamalus* just above a band of white barnacles, *Semibalanus*. Connell came up with four likely causes of this zonation: the physical conditions characteristic of each zone, predation, competition, and where the barnacles were able to settle in the first place. While barnacle larvae float freely in the water,

adult barnacles are permanently cemented to a surface. Because adults are unable to move, their distribution is limited by where larvae settle. Through careful observation, Connell found that larvae of both species had settled well outside the area where he saw the corresponding adults. So, he reasoned, the zonation must be caused by something else.

Next, Connell turned to physical conditions. Perhaps barnacles outside their usual zone die from stresses like temperature extremes or dryness. To test this hypothesis, Connell moved

Above: The two young Semibalanus *barnacles that appear in the lower center of this photograph are tiny now but may overgrow or undercut the older* Chthamalus *barnacle. Top left: Intertidal zonation on Washington's Olympic Peninsula.*

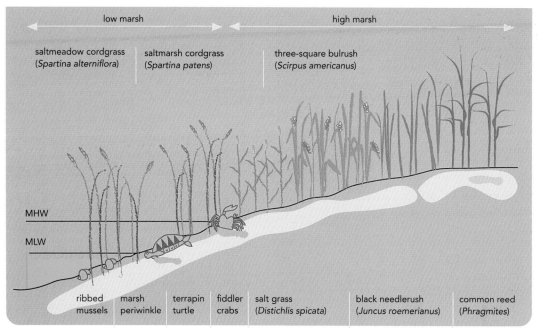

low marsh		high marsh		
saltmeadow cordgrass (*Spartina alterniflora*)	saltmarsh cordgrass (*Spartina patens*)	three-square bulrush (*Scirpus americanus*)		

MHW

MLW

ribbed mussels	marsh periwinkle	terrapin turtle	fiddler crabs	salt grass (*Distichlis spicata*)	black needlerush (*Juncus roemerianus*)	common reed (*Phragmites*)

In salt marshes around the world, various grasses and rushes characterize different levels in the marsh. Each species is adapted to different conditions—more or less time flooded by salt water, for instance. Plants play an important role in shaping the marsh. Spartina alterniflora, *for instance, can form dense peat that resists erosion.*

barnacle-covered rocks to areas high and low in the intertidal zone, and kept track of which species survived where. While both species could survive lower than their normal range, neither could survive higher. Physical stress apparently set the upper limit of both types of barnacles, but not the lower limit.

Perhaps biological interactions kept them from moving lower. There was a possibility that certain predators lived only in particular areas, and one barnacle species only survives where those predators are absent. Or maybe competition was to blame. Adult barnacles cannot move and therefore often compete fiercely for space. Perhaps one species outcompeted the other, restricting its range.

Connell set up two more experiments. He enclosed some groups of barnacles in cages that kept predatory snails away, and left others exposed. He also tested the importance of competition by letting larvae of both species settle in bare patches, then monitoring the interaction of individuals as they grew. While predation did affect the lower limit of *Semibalanus*, the effects of competition were more striking. *Semibalanus* routinely killed *Chthamalus* by growing over and smothering them, or growing under them, prying them off the rock.

THE WRONG SIDE OF THE ROCK

These experiments and others gave rise to a paradigm of intertidal zonation: the upper limits of an organism's zone are usually set by physical factors—heat, light, dryness—and lower limits are set by biological factors like competition and predation. But these factors can interact in even more complex ways, as ecologist Chris Harley showed on Washington State's Olympic Peninsula. On north-facing, wave-exposed rocks, a turflike alga forms a band in the high intertidal zone, safe from herbivores that are unable to spend as much time out of the water. On the hotter south-facing rocks, this alga is absent. The increased light and heat make it impossible for the alga to survive at its usual tidal height, but the height at which the herbivores live is unaffected. The alga, squeezed between certain death above and below, disappears.

Where the Sea Breaks Its Back

The rocky intertidal zone is a world of extremes. Organisms can experience temperature shifts of 40°F (22°C) within 12 hours, and during the same time period go from being completely submerged to completely exposed. In winter, they may become coated in ice, and the force of each wave often exceeds the wind force of hurricanes. Many intertidal animals can feed only when underwater, so while the tide is low, they go hungry. Rain can create pools of freshwater, while on a blazing hot day evaporation can make tide pools extra salty. Intertidal organisms are evolutionarily adapted to these harsh and variable conditions, but there is still a cost.

Despite, or perhaps because of, these conditions, the rocky intertidal zone is dynamic and full of life, a favorite of naturalists and scientists alike. Colorful sea stars, a plethora of algae, and the ever-popular hermit crabs are some of the more visible creatures to reward visitors. What may appear to be lifeless lumps may turn out to be barnacles or snail-like limpets hunkered down waiting for the water's return. Ribbon worms may stretch among the barnacles, waiting to paralyze their prey with a quick strike

Above: Colorful red and purple starfish are found on the rocks of British Columbia's intertidal zone. Top left: Sea otters are a keystone species in the Pacific Northwest, eating urchins and allowing kelp to flourish.

of their poisoned proboscis. Lower down, a few hardy fish may be in tide pools or under rocks. The more one looks, the more one sees.

"THE DUMBEST OF EXPERIMENTS"

Some important science begins with simple observations. On the outer coast of Washington State, ecologist Robert Paine noticed that mussels only occurred above a certain tidal height. Below was a mosaic of bare rock, barnacles, anemones, seaweed, and ochre sea stars. Paine knew that mussels were a favorite food of the ochre sea star, and he wondered whether predation by this star kept mussels from living lower down.

To test his hypothesis, Paine did what he called "the dumbest of experiments." He removed all the ochre stars from some areas, and left others alone. After three years, the bottom of the mussel zone was several feet lower in areas where he had removed the stars, but relatively

When dominant organisms like mussels are removed, organisms like the sea sacs, above, may quickly colonize the free space.

unchanged where they had been allowed to remain. His hypothesis was supported!

However, Paine noticed something else: without stars, what had been a diverse community now looked like a mussel monoculture. While closer inspection revealed that mussel beds do harbor some diversity, the absence of sea stars had nonetheless completely reshaped the community. Paine coined the term "keystone predator" to describe species, like the ochre star, whose influence on ecosystem structure is much greater than the number of individuals of that species would suggest. Sea otters are another keystone predator. Otters eat sea urchins, urchins eat kelp, and kelp forests harbor an amazing diversity of fishes and invertebrates. In the presence of otters, urchin populations are kept under control, and kelp forests flourish. When sea otters along the west coast of North America were hunted to near extinction around the end of the nineteenth century, urchin populations exploded and kelp forests disappeared.

Paine's experiment also reveals the importance of indirect effects, underlining the difficulty of predicting what will happen based on prior knowledge of who eats what or who competes with whom. While sea stars do not interact with anemones directly, they affect anemone abundance by eating mussels, which are so good at taking over space

Children should be reminded not to disturb fragile ecosystems when beachcombing.

INTERTIDAL ETIQUETTE

1) If you turn over a rock to see what lies underneath, gently replace it the way it was when you are done. Organisms from the underside may die if left exposed, while those from the top will be suffocated if left underneath.
2) Step lightly. Not all animals are obvious, and the effects of people walking through the intertidal zone can create lasting damage.
3) Put animals back where you found them. Animals survive best in particular locations.
4) Be careful what you bring home with you. It can be hard to tell what is alive and what is not, and even casual collection by beachgoers can have negative effects on intertidal populations.

that without predators they can all but eliminate some species of anemone. Profound new understandings can sometimes arise from good observational skills and a simple, "dumb" experiment.

Pasture of the Sea

Compared to colorful reefs or the wave-swept rocky intertidal zone, sea-grass beds may not seem like much. When the water is out, the narrow blades lie plastered limply against the sand or rock. But underwater, sea-grass beds teem with life, great and small. Along the northwest coast of the United States, for instance, nearly 200 invertebrate species live in eelgrass beds, and 70 fish species use the beds for food, shelter, and reproduction, including salmon and herring. In the tropics, sea-grass beds are home to several threatened or endangered animals, including manatees, seahorses, and sea turtles.

UNDERWATER PLANTS

Sea grasses are remarkable in part because they are flowering plants that live in the ocean. Like land plants, but unlike seaweed, sea grasses have roots for taking up nutrients, and produce flowers, pollen, fruits, and seeds. While land plants typically use wind or animals to pollinate their flowers and disperse their seeds, water currents play this role for sea grasses. Seeds may travel individually, often with a trapped air bubble aiding in buoyancy, or waft about still attached to broken or uprooted stems. Most seeds land fairly close to the bed from which they come, but some may settle successfully over 80 miles (130 km) from their origin.

Although capable of sexual reproduction, most sea grasses reproduce vegetatively, with rootlike rhizomes extending out from existing plants and sending up new blades. Thus a sea-grass bed that appears to be made up of thousands of individuals may really contain just a handful of clones, some of which may be over a thousand years old.

HOME TO MILLIONS

The richness of sea-grass communities rests on the plants' ability to provide food, substrate, and shelter for a wide variety

Above: Manatees, or sea cows, eat a variety of food but rely heavily on sea grasses. A large manatee can eat over a hundred pounds of vegetation in a day. Top left: Eelgrass near the surface at low tide.

of organisms. A dense carpet of bacteria and single- and multicellular algae growing on each blade provides food for microscopic grazers called flagellates. These are eaten by tiny harpacticoid copepods, relatives of shrimp and crabs, which in turn are an important food source for juvenile salmon and other carnivores. Hosting this food web comes at a price: Each blade may carry up to two times its own weight in other living things and must grow quickly to maintain any uncovered surface with which it can carry out photosynthesis. Photosynthesis is important to sea grasses not only as a source of energy, but also because air bubbles that are produced during photosynthesis help their blades to stay afloat. Dense sea-grass forests slow

SEA GRASS FACTS

Globally, there are about 50 species of sea grasses, including eelgrass, tape grass, surf grass, and turtle grass. In Alaska's Izembek Lagoon, brant geese eat 2,000 tons of eelgrass annually before flying 3,000 miles to Mexico, where they spend the winter. By some estimates, sea-grass plants shed and replace their leaves seven times a year. This means that sea-grass beds produce three billion pounds of nutrient-rich detritus annually. Relative to historical levels, sea-grass cover has declined by half in Florida's Tampa Bay, 75 percent in the Mississippi Sound, and 90 percent in Galveston Bay, Texas.

The color of the eelgrass isopod often matches its diet, often kelp or eelgrass.

currents and provide hiding places, making them common breeding and nursery grounds for animals including herring, salmon, and Dungeness crabs. Eelgrass blades can be coated with billions of herring eggs, providing a feast for animals from birds and bears to sea stars and fish. A decline in scallops in Waquoit Bay, Massachusetts, followed a decline in eelgrass, and scientists hypothesized that the drop in scallop numbers was due to the lack of eelgrass for young scallops to use as transitional habitat between their life as floating larvae and benthic adults.

UNDER PRESSURE

Despite their importance, sea-grass beds are under threat from many sources. Because sea grass needs strong sunlight to survive, docks and other over-water structures kill sea grass both underneath the structure and in a "shadow zone" on either side. Increased coastal development and industry generate pollution and sedimentation that can kill or stress sea grass. Nonnative species and climate change are also taking a toll. A less common but potentially catastrophic problem is disease: In 1931 and 1932, a mysterious wasting disease killed 90 percent of the sea grass on both sides of the North Atlantic. The populations have yet to recover.

Seahorses are weak swimmers and hold on to sea grasses with their tails to avoid drifting with the current.

Rivers and Tides

Salt marshes, which grow where temperate rivers and streams meet the sea, are places of transition, subject to the rhythms of both river and tide. Like all intertidal habitats, the marshes have a gradient of tidal flooding. The high marsh, which is covered with water only during storms or the highest tides, appears almost terrestrial, but the organisms living there must be able to tolerate the occasional saltwater flood. Unlike most marine ecosystems, marshes boast a diversity of true plants rather than just algae. Grasses, sedges, and rushes give them a pastoral feel, while sea lavender adds a floral touch. Come fall, the humble pickleweed turns a brilliant red, the marshes' answer to the maples of New England.

Fed by nutrients from land and sea, salt marshes have among the highest mass of life per unit area of any marine community. The rich plant life supports a similarly rich abundance of ani-mals. Hermit crabs and worms burrow in the mud, feeding on decaying organic matter. Clams and mussels bury themselves, too, but feed instead on the microscopic algae and animals carried by the water. Raccoons hunt for crabs and fish while mice graze on the greenery. Fish large and small swim the waters, and resident and migratory birds join the feast as well.

GO WITH THE FLOW
One of the unique elements of estuarine life is that water often moves toward the ocean on the surface, where river flow dominates, and in the reverse direction on the bottom, where salt water dominates. This bidirectional flow allows even weakly swimming organisms to have some control over where they go. By swimming up, organisms can move out of the estuary, and by swimming down, they can catch an inbound current. Crab larvae, which spend weeks or months in the water before settling down as adults, are known to behave in this fashion. At different times of day or at different life stages, larvae will change whether they swim up or down, and in so doing determine

Above: A diversity of birds depends on marshes for feeding and breeding. Some eat plants, while others, like these egrets, feed on the rich supply of fish and invertebrates. Top left: A marsh in Namibia.

whether they are carried into the estuary or out of it. Not bad for an animal with a brain smaller than a pinhead!

WATERLOGGED WEALTH

Beyond their natural appeal, salt marshes provide a variety of tangible benefits for humans. Many commercially important fishes—including salmon, herring, and halibut—spend at least part of their lives in estuarine wetlands, taking advantage of the relative calm and plentiful food. Recreational and commercial fishing of crabs, oysters, and other invertebrates is common in marshes as well. Wetlands can act as water filtration systems and buffers against flooding. No wonder ecologist Edward Maltby referred to wetlands as "among the Earth's greatest natural assets . . . mankind's water-logged wealth."

Yet this wealth is under threat. Accelerating sea-level rise as a result of climate change threatens estuaries with drowning. Dams reduce and control water flow, starving wetlands of sediments and changing the balance of salt and fresh water. High levels of pollution are common, with sources ranging from factories and mega-farms to leaky septic systems and people dumping used motor oil into sewers. Nonnative species are a problem as well. For instance, nonnative smooth cordgrass is threatening native mudflat communities in Washington State's Willapa Bay, transforming what is now excellent oyster habitat into a grassy marsh. In the cordgrass's native Northeast, meanwhile, the grass itself is threatened by the aggressive growth of common reed. While the reed is probably a native, something in the reed, the environment, or both has transformed it into an invasive plant that is radically reshaping coastlines.

A split-hull hopper dredge releases sediment sucked from the bottom of the Mississippi River. Such sediment once replenished wetlands along the Gulf Coast but now is dredged and dumped offshore in the Gulf of Mexico.

LOST PROTECTION

Development and alteration of natural river flow have destroyed millions of acres of once-extensive wetlands along the Gulf Coast of the United States. Healthy wetlands act like sponges, absorbing excess water and releasing it slowly later, and scientists warned that wetland loss made the Gulf Coast more vulnerable to flooding and storm

Salt marshes provide calm and beauty in good weather, and protection during storms. Some are used to treat sewage.

surges. In 1998, after Hurricane Georges nearly hit New Orleans, and again after Hurricane Ivan in 2004, Louisianans asked the federal government for help restoring coastal wetlands. The government balked, causing gubernatorial aide Sidney Coffee to wonder, "What is it going to take for Congress and the president to realize this is not just another project? Would we have had to get hit by the big one? Who wants to wait for that? Surely it shouldn't have to take loss of life?" Hurricane Katrina proved scientists' predictions were catastrophically correct, and sent a powerful message that wetland restoration is not just another project.

Forest by the Sea

Ask most people about tropical marine ecosystems, and they think of coral reefs. Ask about tropical forests, and they think of rain forests. But most tropical coastlines are rimmed with an entirely different ecosystem—mangrove forests. Full of mosquitoes and difficult to traverse, they are not popular with tourists. Still, they do possess a certain grandeur. Klause Rützler and Ilka Feller, ecologists at the Smithsonian Institution, captured the majesty of these habitats in their writings on Caribbean mangrove swamps: "One perceives a forest of jagged, gnarled trees protruding from the surface of the sea, roots anchored in deep, black, foul-smelling mud, verdant crowns arching toward a blazing Sun . . . Here is where land and sea intertwine, where the line dividing ocean and continent blurs."

Above: Mangrove roots dangling above the bottom. The roots look fuzzy due to the density of organisms growing on their surface. Top left: A river winds through Australian mangroves.

THE MANGROVES

Mangroves are trees adapted for life at muddy edges of tropical and subtropical seas. Although there are just 50 to 80 species, mangroves come from around 20 different families and range in form from mere shrubs to trees tens of meters tall. Different species live in different habitats, from sheltered coasts to extremely salty mudflats to estuaries. The mud in which mangroves grow is often unstable and low in oxygen, and mangroves have evolved a variety of types of aerial roots. These stabilize the tree, and many have "breathing" pores that take up oxygen from the air when the tide is out.

Although mangroves live in salty environments, salt could kill the trees if they had no way to get rid of it. Some mangroves prevent the salt from entering in the first place by filtering the water they take in through their roots. They do this so well that cutting open the root of a salt-excluding species produces water fresh enough to drink. Other species take up salt water but excrete the salt from their leaves, roots, or branches.

THE COMMUNITY

Mangrove forests support their own, unique community of plants and animals. Fallen leaves and branches provide nutrients, and living trees stabilize the environment by trapping sediment and preventing erosion. Algae and animals live on and around the trees, and at high tide hundreds of species of fish, shrimp, and crab swim among the roots. Mangrove forests even support ecosystems beyond their borders by reducing the flow of

A dense tangle of roots helps mangroves stabilize sediments and get enough oxygen. They also create channels that can be safe havens where inhabitants can ride out storms.

sediments and pollutants from land to sea. In this way, they enhance the growth of adjacent sea-grass meadows and coral reefs. Many fish species that live on coral reefs as adults spend their early life safely among the roots of mangrove trees. Mangroves also support human communities, providing food, fuel, medicine, and building materials for coastal peoples around the world.

Unfortunately, mangrove forests are disappearing. An estimated 50 percent of mangroves have been lost globally. They are cut down to create beaches for tourists or to make way for shrimp farms. As the human population grows, subsistence use that was once sustainable takes a toll. Climate change is also a threat, as rising seas can drown mangrove forests and changes in freshwater input can make currently forested areas unsuitable for mangrove growth.

PROTECTION FROM THE WAVES

In a stark example of the importance of mangroves, consider the fate of two Sri Lankan villages during the December 2004 tsunami in Asia. In Kapuhenwala, a village surrounded by acres of healthy mangrove forest, just two people died. Less than 6 miles (10 km) away was the village of Wanduruppa, where mangrove forests had been severely degraded. Here, between 5,000 and 6,000 people died. In 1999, a supercyclone hit Orissa, India, killing over 10,000 people and washing away entire villages. Meanwhile, around the Bhitarkanika Wildlife Sanctuary, which contains the second largest mangrove forest in India, villages and lives were spared.

Mudskippers have tear ducts to keep their eyes moist during protracted time in air and can see better in air than water.

A FISH OUT OF WATER

Most fish do not voluntarily spend much time out of the water, but the bug-eyed mudskippers common in mangrove forests are fish of a different sort. They can "breathe" through their skin as long as it stays moist, and "hold their breath" by filling their mouth and gill chamber with water before venturing onto land. Their front fins act almost like legs, allowing mudskippers to "walk" and even climb trees! Male mudskippers engage in fierce ritual battles for prime patches of mud.

Although archer fish stay in the water, they make meals of insects and other small terrestrial animals. If the prey is less than 12 inches (30 cm) above the water, the fish may leap out and grab it directly. For more remote prey, archer fish turn into submarine water cannons, accurately targeting insects more than 3 feet (1 m) away with forceful jets of water they shoot out of their mouths, knocking the prey into the water where it can be easily gobbled up.

Cycles Connecting Land and Sea

Scientists often divide the world into different realms—land, sea, and freshwater, for instance, or the solid earth and the atmosphere above—but boundaries between these realms are not absolute. Molecules constantly cycle between the living and non-living, between air, water, rock, and soil. Many nutrients that fuel intertidal organisms have been carried to the ocean from thousands of miles inland, and nutrients from the ocean may cycle back into terrestrial and freshwater habitats through a variety of pathways.

Some of the most important nutrients are nitrogen, phosphorus, iron, and silica. Nitrogen is used to make amino acids, which are the building blocks of proteins, and chlorophyll, the main source of energy for photosynthetic organisms.

Phosphorus is an essential part of DNA, RNA, and cell membranes, and many animals use calcium phosphate to make bones, teeth, or shells.

Nitrogen can occur as a gas, as part of living things, or dissolved in water, where it commonly occurs in three forms: nitrate, nitrite, and ammonium. Although nitrogen makes up 78 percent of Earth's atmosphere and 48 percent of dissolved gases in seawater, most multicellular organisms are not able to take up nitrogen gas. They depend on an array of microorganisms capable of converting nitrogen into more usable forms. Some bacteria "fix" nitrogen by combining it with hydrogen to make ammonium. Others combine ammonia with oxygen to make nitrite, while still others convert the nitrite to nitrate, the nitrogen-containing compound most readily used by green plants.

Above left: Deer pea, common in low-salinity coastal marshes, hosts root bacteria that are important for nitrogen cycling. Above right: Sewer outfalls like this can affect local nutrient cycles. Top left: Salt flats and marshland with oil tanks and waste from phosphate production.

SOURCES AND CYCLES

Nutrients enter the marine environment from a number of sources, including rivers, rain, dust, sewage outfall, and waste dumped by boats. Nitrogen and phosphorus used to fertilize fields far from the ocean and immense quantities of waste from commercial hog and cattle operations find their way into streams and rivers, where they are carried to the ocean. Burning fossil fuel also releases nitrogen into the environment. Likewise, the weathering of bedrock releases significant amounts of phosphate that can make its way through rivers to the ocean.

Algae and other photosynthetic organisms take up nitrogen, phosphorus, and other nutrients from the water around them. Herbivores in turn take up nutrients from the primary producers they eat, and carnivores from the animals they eat. Living organisms return many nutrients to their environment as waste products, which may be taken up by other organisms, sink to the bottom to become part of sediments and eventually rocks, or rise and become part of the atmosphere. When an organism dies, decomposers, such as microbes and fungi, dismantle its body into a series of building blocks: carbon dioxide, nutrients, water.

There are three phosphorous cycles in the marine environment. One is in the upper layer of the ocean known as the mixed layer, where phosphorus cycles rapidly through living things via feeding, death, and decomposition. A somewhat slower cycle occurs when organisms or phosphate-rich organic matter falls deeper into the ocean. The phosphorus they carry circulates in the deep ocean and may return to surface waters only after hundreds of years. Some phosphorus sinks all the way to the ocean floor, becoming buried in sediments and contributing to the longest of the phosphorus cycles. Once buried, phosphorus can remain out of the loop for millions of years, released only through volcanic activity or when geological processes uplift the phosphate-rich rocks they form to where they can be mined by humans for fertilizer, or leached out of rocks by natural processes.

Sockeye salmon turn a river red in Alaska's Katmai National Park as they migrate upriver to spawn. These salmon will feed bears, eagles, invertebrates, and countless other animals, forming a key link between land and sea.

FISH AS FERTILIZER

While most people think of salmon as food, business, or an icon of the Pacific Northwest, some ecologists think of them as nutrient-transfer machines. Born in freshwater, salmon swim hundreds of miles to grow and mature in the ocean before returning to their birthplace to start the next generation. When they head inland to spawn, they bring more with them than a desire to reproduce. They bring phosphorus and nitrogen from the sea. Because most salmon die immediately after spawning, their bodies release these nutrients to the ecosystems around their spawning streams. Scientists can measure the signature of marine nutrients in plants and animals in streams and forests far distant from the shore. Thus the loss or decline of salmon may translate into less fertilizer for the areas where they once spawned.

CHAPTER 10

LIFE IN THE OPEN SEA

Left: A blue-water diver descending. She is carrying a glass jar, which is used in the collection of gelatinous specimens. Upon returning to the surface, the specimen will be studied in the ship's laboratory. Top: A humpback whale breaching. Scientists are not sure exactly why whales breach. It could be a form of communication; it could also be a way for the whale to groom itself. Bottom: Jellyfish.

The open ocean is different from the familiar terrestrial world in a number of fundamental ways, with some interesting and surprising consequences. It is three-dimensional to an extent unknown on land. Many marine animals routinely travel vertical distances of thousands of feet every day, sleeping and feeding suspended in a liquid medium. A wide range of organisms inhabit the open ocean, from the smallest bacteria to the largest animals ever to live on Earth. Nearly all of the primary producers, which power the entire oceanic food web, are microscopic.

To thoroughly study the organisms of the open ocean demands a variety of approaches. To see what mammals, birds, and fish do when they dive beneath the surface, scientists equip individual animals with tiny computers that record time and depth, or, in the case of some seals, an underwater video camera. To study the smallest creatures, oceanographers drag nets through the water and look at their haul through microscopes or using molecular techniques. Researchers do what is called blue-water diving to get a sense for larger but more delicate inhabitants of the realm, the jellies. Because there are no landmarks and disorientation is all too easy, blue-water divers attach themselves to a rope tied to their ship before descending into the open water.

Swimming in Molasses

Picture the following scenario: Pour cream into a nice hot cup of coffee, and the cream stays in a little blob, right where it was poured in. Stir the coffee several times in a clockwise direction, slowly dispersing the cream. Reverse the stirring direction, though, and amazingly the cream stirs right back into its original ball! Such is the world of the smallest marine organisms.

Objects in motion tend to stay in motion, and objects at rest tend to stay at rest. At high Reynolds numbers, inertial forces dominate and flow is turbulent. Humans generally live at high Reynolds numbers. When someone is riding a bicycle and stops pedaling, they will coast for a while before stopping. This is an example of inertia in action.

Viscosity is a measure of how strongly a fluid resists flow. Air, which offers very little resistance to flow, has a low viscosity, while honey has a higher viscosity. At low Reynolds numbers, viscous forces dominate and flow tends to be smooth. Bacteria live at low Reynolds numbers. A marine bacterium moving along at typical speeds coasts just a trillionth of a centimeter once

MOVING IN THE HERE AND NOW

In the language of fluid dynamics, a key determinant of how planktonic creatures experience the world is the Reynolds number at which they live. This is the ratio of two forces, inertial and viscous, that determines whether water flowing around the organism is smooth or turbulent.

Inertia is the concept described in Newton's first law of motion:

Above: Large, fast-moving animals like killer whales experience significant turbulence in their lives. Microscopic organisms living in the same environment experience virtually none. Top left: Honey is a viscous substance. To survive in an environment as viscous as honey, an organism would have to function efficiently in slow motion.

it stops swimming. At low Reynolds numbers, objects in motion have almost no tendency at all to stay in motion. The smaller and slower the organism, the lower its Reynolds number.

BOUNDARY LAYERS

What is life like at low Reynolds numbers? It has often been compared to swimming in molasses while moving no part of one's body faster than the hands of a clock. At low Reynolds numbers, the water surrounding a particle becomes almost an extension of that particle, forming what is called a boundary layer. Reach out to grab a ball in a molasses-filled pool, and the ball moves away as a hand pushes against the layer of molasses surrounding the ball. And, as alluded to in the coffee example, flow is completely reversible—pull the hand away from the ball, and the ball moves back to its previous position. For very small animals to move or capture food particles, they must move their appendages in ways that do not simply push water back and forth.

WETTER AND DENSER

Some aspects of life in the ocean are unthinkable on land even for organisms living at higher Reynolds numbers. For one thing,

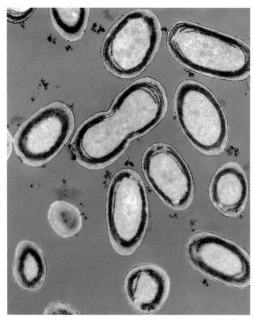

A colored electron micrograph of single-celled Prochlorococcus *plankton, which have no mobility and drift where ocean currents take them.*

water is wet. Except for those that live in the intertidal zone, marine organisms do not need fancy ways of keeping themselves moist. This allows the existence of free-living embryos and other tiny organisms, and allows many animals to put their breathing apparatus on the outside rather than inside their bodies.

Water is also denser than air, which makes floating easier. No living thing is lighter than air, but being less dense than water is easy. Planktonic organisms are often equipped with air bladders or oil droplets that make them buoyant. Some, like jellyfish, make their entire body from a material close to the density of seawater. While birds spend energy to keep themselves aloft, most planktonic organisms do not.

A transparent squid hovering nearly 3,000 feet beneath the surface of the ocean.

INVISIBLE ORGANISMS

Few terrestrial animals have achieved transparency, but see-through bodies are common in the plankton. There are transparent comb jellies, jellyfish, round worms, tunicates, segmented worms, arrow worms, crustaceans, snails, and fish. In part, transparency may be a form of camouflage in the open ocean. In response to living in a world lacking surfaces with which to blend in, transparency is the only possible disguise. It may also be that a planktonic lifestyle makes it easier to achieve transparency. The buoyancy of water reduces the need for rigid skeletal supports, which are often opaque, and floating animals can build bodies that are mostly water. Being beyond the well-lit surface zone greatly reduces the need for dark pigment as protection from harmful ultraviolet radiation. Still, the presence of intensely colored animals in the deep ocean is a reminder that many mysteries remain unsolved.

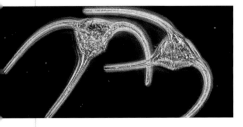

Tiny Drifters

Most people find it easier to relate to things they can see, yet most of the ocean's biodiversity is invisible to the naked eye. Seawater teems with microscopic algae, animals, bacteria, and protists. A bucket that appears to hold only water can harbor all the dramas of life: the hope of youth, relentless hunting and killing, and reproduction. The tiniest sea creatures are collectively called *plankton*, a term for organisms unable to swim against a current. Some cannot swim at all, while others swim only weakly. There are even some larger animals, such as jellyfish, whose seemingly aimless movement qualifies them to be categorized as plankton.

BLOOMING OCEANS

At the base of the planktonic food web are the phytoplankton, which get their energy from sunlight. Most are single-celled, though they may form chains or clumps. While individuals are generally invisible to the naked eye, under the right conditions phytoplankton reproduce like mad, forming extensive blooms

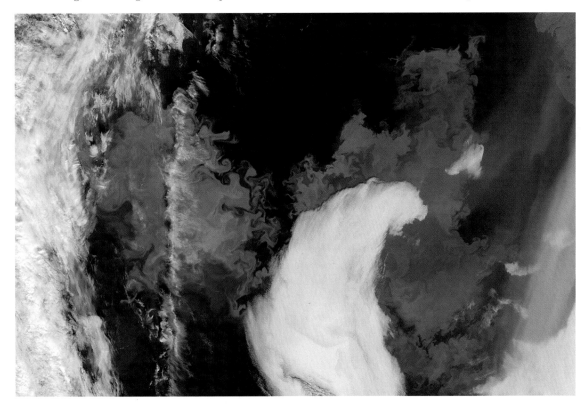

Above: This satellite image of the Barents Sea shows a distinct blue-green phytoplankton bloom, indicating a surge of reproduction in the sea's plankton community. Top left: Ceratium tripos is a dinoflagellate. These organisms can move independently, consume other organisms, and are skilled at photosynthesis.

that discolor the ocean.

The most common phyto-plankton are diatoms, called golden algae, which come in an array of shapes, from triangles to hatboxes to spiky spheres. Their hard, porous capsule of silica, the chemical that forms glass, allows light to pass through but provides some protection. Dinoflagellates are a common type of phytoplankton that defy easy categorization. Using a pair of flagella, they can move as animals do and often consume organic matter, even attacking living creatures. Yet they are also very efficient at photosyn-thesis. Even more puzzling, a single species may take 20 or more different forms, ranging from a more typical form with a rigid shell and two flagella to a shapeless amoeba. Some dinoflagellates can cause lethal red tides.

CONSUMER SOCIETY

The ocean is also home to hungry hordes of microscopic animals, the zooplankton. Almost every animal phylum has at least a few representa-tives in the plankton, either as permanent members of the community or during particular stages of their life cycle. Most follow the motto "Live fast, die young," in many cases cycling through several generations a year. Some eat bacteria and phytoplankton, while others prey on other zooplankton. The hunt can be dramatic: A planktonic snail called the sea angel has jaws on top of its head that shoot out to snatch its prey in a decidedly unangelic fashion.

Despite their small size, zooplankton are critical members of marine food webs. Blue whales, the biggest animals of all, rely on plankton for their food, as do penguins, whale sharks, and many other ani-mals. Particularly important are the crustaceans known as copepods and krill, which are both abundant and fatty. Declining krill populations near breeding grounds may be linked to the decline of some penguin popula-tions.

Craig Venter pioneered techniques that helped decode the human genome. He now uses these techniques to explore the ocean's microbial diversity and has found hundreds of new species.

The quantity and variety of organisms that make up the plankton change seasonally, as well as on a daily or even hourly basis as currents and tides mix things up. Phytoplankton numbers rise and fall with available light and nutrients. Because of their short generation time, zooplankton populations rapidly swell following a phyto-plankton bloom, then fall off as phytoplankton supply diminishes. Other members of the plankton appear seasonally, based on the reproductive cycles of benthic animals with planktonic larvae. Visit a coral reef after a full Moon, and the water may be swarming with coral larvae. At temperate latitudes, spring brings millions of larval sea urchins, mussels, and other animals.

Amphipods such as these are a food source for many animals, including blue whales.

Carbon Cycles

Carbon, an essential element of all life on Earth, is a common component of the nonliving world as well. Carbon bonds in organic molecules store energy in cells, and the carbon bonds in sugar provide a good many of its calories. The ocean, atmosphere, and terrestrial systems all serve as a reservoir for carbon, taking it up and releasing it over time. The ocean stores the greatest amount of carbon, and the atmosphere the least. Carbon also ends up in rocks, where it may remain trapped for millions upon millions of years. The cycle of carbon through the living and nonliving, the solid and gaseous, forms the largest of all biogeochemical cycles. The carbon cycle influences life on Earth directly as a nutrient, and also indirectly through its role in determining Earth's climate.

SOURCES

Many living organisms, including plants and animals, release carbon in the form of carbon dioxide as a by-product of respiration. Forest fires release carbon in gaseous form to the atmosphere and as ash and charcoal to the soil. Animals excrete about 10 percent of the carbon they eat in their wastes, which in water creates dissolved organic carbon, or DOC.

On a longer time scale, volcanic eruptions bring up carbon from deep within Earth, releasing it into the atmosphere or depositing it in ash and lava. The burning of fossil fuels like oil and gas likewise releases carbon that has been stored beneath the ground for eons. Carbon-rich sediments or rocks buried on the seafloor may be uplifted by tectonic movement or exposed by falling seas. Once exposed, weathering of rocks and sediment make the carbon available again.

SINKS

What happens to carbon released by these sources? Algae and other photosynthetic organisms take up carbon dioxide and use it to make sugars and carbohydrates. Bacteria and other planktonic organisms take up DOC from the water around them. Animals take up carbon when they feed, storing

Above: The white cliffs of Dover are composed of lime-stone sediments formed by the accumulation of skeletons and shells over hundreds of millions of years. Top left: Phytoplankton blooms, as seen in this Sea-viewing Wide Field-of-view Sensor (SeaWiFS) image, release carbon into the ocean.

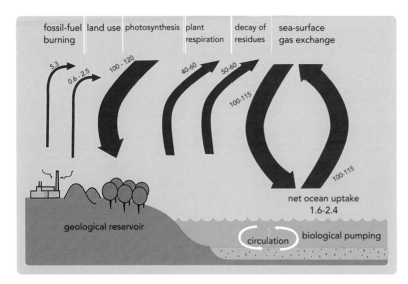

fossil-fuel burning	land use	photosynthesis	plant respiration	decay of residues	sea-surface gas exchange

5.3
0.6 - 2.5
100 - 120
40-60
50-60
100-115
100-115

geological reservoir

net ocean uptake
1.6-2.4

circulation   biological pumping

Top: This diagram illustrates the various components of the carbon cycle. Bottom: A ledge of foraminiferan limestone is visible in the upper left corner of this image. The remains of undersea wildlife can metamorphose over time into a variety of substances, including limestone, coal, oil, or natural gas.

also buried in sediments on the ocean floor, which may metamorphose into limestone. Under the right conditions, dead plants and animals can form coal, oil, or natural gas, which removes carbon from the carbon cycle for millions of years.

FERTILIZING THE OCEAN

Because of carbon dioxide's role in global warming, many people are looking for ways to reduce the amount of it in the atmosphere. One approach is to increase the uptake of carbon by carbon sinks, which are reservoirs that take up released carbon from other parts of the carbon cycle, such as growing vegetation and the ocean. On land, many organizations, businesses, and governments are planting or protecting trees for just this purpose. Perhaps the open ocean could be used in a similar fashion. While it is hard to "plant" phytoplankton, its growth can be increased by adding fertilizer. In the open ocean, phytoplankton growth is usually limited by iron, and scientists have experimented with dumping large quantities of iron far out at sea. As expected, the iron stimulated phytoplankton growth, which would result in the extraction of more carbon from the atmosphere. Critics point out that widespread open ocean fertilization would profoundly alter pelagic communities, but would not make enough of a dent in atmospheric carbon dioxide to offset the costs.

it in their bodies for the duration of their lives. Many animals also use carbon in their shells or skeletons. When these animals die, their hard parts may be consumed by other creatures, decomposed in the soil, or dissolved into the water. Shells and skeletons are

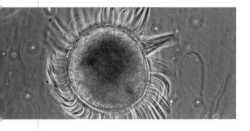

From Egg to Egg

From an evolutionary perspective, the survival of offspring and future generations is the only real measure of success, and marine animals display a marvelous diversity of approaches to reproduction. Mammals, birds, snails, sea slugs, cephalopods, and crustaceans all have internal fertilization, in which sperm is transferred directly from one individual to another. A few fish species also fertilize eggs internally, but most fishes, along with echinoderms, bivalves, and jellyfish, have external fertilization. In some cases, fertilization is still a somewhat intimate affair: A pair of fish, for instance, may do a mating dance that culminates in simultaneous release of gametes. More often, animals with external fertilization engage in a behavior known as free-spawning, in which sperm and eggs are simply expelled into the water.

THE PULL OF THE MOON

Free-spawning is a tricky affair. Sperm do not last long once released, and finding a receptive egg can be difficult. Thus free-spawners have evolved a variety of mechanisms that increase the chance of successful fertilization. Many, like palolo worms and other polychaetes, time the release of sperm and eggs to the lunar cycle, so that many individuals will spawn at the same time. These worms also swim to the surface to spawn, increasing the proximity of reproducing individuals. While mature corals are fixed to the bottom, their eggs float. Corals in a given reef synchronize their spawning, and sperm and eggs may form extensive slicks on the surface after a reef-wide spawning event. Because fertilization success can be low for free-spawners, they release thousands, millions, or in some cases billions of eggs each year. Even if less than 1 percent of an individual's eggs are fertilized, hundreds of embryos will still be produced.

Not all animals reproduce sexually. Aggregating anemones split in two, with each half of the anemone setting off in a different direction. Sponges can reproduce new individuals from even a small piece of an existing animal. Colonial invertebrates like hydroids and many tunicates simply bud new individuals off of existing ones.

PARENTAL CARE

Some animals invest a lot of time and energy into each offspring. Orcas usually produce one calf every five years or so, which the mother nurses for up to two years. Others, like purple sea urchins, produce millions of eggs annually, each of which

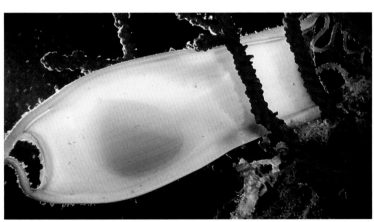

Above: A dogfish egg case secured to a sea fan to protect it from predators. Each dogfish egg case contains a single embryo, which matures into a young shark in about nine months. Top left: The planktonic larva of the bristle worm, Eulalia viridis, *is ringed by bands of cilia used for feeding and movement.*

has enough yolk to last the developing embryo just the few days it takes to develop the ability to feed on its own. These free-spawners never know the fate of their offspring. Still other animals take an intermediate route. Sharks and their kin provide developing embryos with rich egg yolks that look surprisingly like those of a chicken, and many invertebrates likewise produce eggs with generous amounts of yolk. Some animals, like the six-rayed sea star, brood their young in or on their own bodies until the offspring are capable of independent living.

A few invertebrates have a particularly gruesome way of provisioning their offspring: Females produce capsules containing many embryos, and the fastest-growing embryos eat their siblings. In one species of snail, the siblings start eating each other before they even develop a stomach!

While some larvae settle down near their parents, others may be carried thousands of miles before choosing a spot to live. Larvae produced by mussels on the coast of North America, for instance, may occasionally make it all the way to Europe.

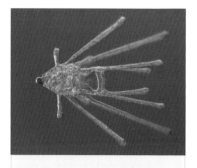

A sea urchin larva, which uses ciliated arms to swim.

BECOMING A GROWN-UP

Larvae and adults of the same species do not always live in the same place. Many benthic animals have planktonic larvae, and some animals split their lives between fresh and salt water. In these cases, the metamorphosis from larva to adult involves a complete change of habitat and is often abrupt. For instance, sea urchins metamorphose from swimming larvae with bilateral symmetry and eight arms to globular adults in just a few hours. Dr. Renae Brodie spent time at the Smithsonian Tropical Research Institute's marine lab on Naos Island, Panama, studying animals with an even more extreme transition. The terrestrial hermit crabs (*Coenobita compressus*) that she examined have marine larvae. At certain times of the year, adults march en masse from the Panamanian forest to the sea to release their young. After weeks in the plankton, the individual crabs metamorphose from larvae that breathe water, float, and feed in the rich, wet ocean to adults that breathe air, walk, and scavenge for food on land.

Coral spawning off western Australia. Coral generally reproduce sexually, releasing gametes into the water, which merge to form larvae. They can also reproduce asexually, when pieces of coral, broken and moved by currents, settle and continue growing in a new location.

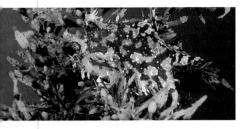

Sea
of Lost Ships

On a December afternoon in 1945, five Navy bombers took off from the Naval Air Station in Fort Lauderdale, Florida, on an overwater navigational training flight. The weather conditions, light winds, and scattered showers were average for a training flight. The instructor was well qualified, and the trainees had the appropriate background. There was no reason to think that this would be anything other than a routine training session. Then, about two hours into the flight, a worried radio message came through—the planes' compasses were malfunctioning and the group had lost their way. This was the last communication received. The planes and pilots were never seen again, and five days of searching turned up no wreckage. The location of this mysterious disappearance was the infamous Bermuda Triangle, a region of ocean credited with all sorts of unexplained events, primarily the disappearance of boats, planes, and people.

The area of ocean in which the Bermuda Triangle sits, the Sargasso Sea, has been accumulating such legends since at least the 1800s. The sea is two million square miles (5.1 million square kilometers) of seaweed-covered calm in the middle of the North Atlantic gyre. Encircled by four strong currents—the Gulf Stream, the North Atlantic Current, the Canary Current, and the North Atlantic Equatorial Current—it acts as an oceanic graveyard. The dearth of wind and current in the area means that anything that floats in is unlikely to leave, including derelict or abandoned ships. In the days of wind-powered ocean transport, many a seafaring crew, including those in convoy, cursed the calm.

AN ALGAL NAMESAKE

The Sargasso Sea gets its name from the floating algae that are so common there. While some species of sargassum, or gulfweed, live attached to the bottom like most big seaweeds, the two species of gulfweed that

Above: Some species of sargassum weed never root in the seafloor. These species spend their entire life floating on currents, held up by small, spherical sacs filled with carbon dioxide. Top left: Sargassum fish live among sargassum weed, wearing fleshy saclike appendages that help them blend into their habitat.

dominate the Sargasso Sea are truly planktonic, completing their life cycles without ever touching down. A multitude of small invertebrates live on the seaweed, and small fish hide amid its leafy fronds, feeding and hiding from predators. Predators also make use of these dense mats of kelp. Some merely pass through looking for dinner, while others are permanent residents, camouflaging themselves and ambushing the unwary. All these animals fertilize the Sargassum with their waste, providing nutrients in an otherwise nutrient-poor region and facilitating the growth of the algae that harbors them.

OPEN-OCEAN NURSERY

Portuguese man-of-wars, jellyfishlike animals with an air bladder protruding above the water and extensive stinging tentacles hanging down below, also accumulate in the

The tentacles of the Portuguese man-of-war both sting potential prey and shelter certain species of fish.

THE GARBAGE PATCH

The same atmospheric and oceanic forces behind the North Atlantic gyre and the calm of the Sargasso Sea create a similar phenomenon in the North Pacific, where strong currents circle clockwise around a calm area the size

Black-footed albatross eating waterborne garbage. The concentration of refuse in the calm waters of the North Pacific Garbage Patch is a deadly anomaly for the area's inhabitants.

of Texas. Here the ocean is covered with something other than seaweed—it is inundated with garbage. Plastic bottles, flip-flops, and other non-biodegradable trash from around the North Pacific end up here, leading many to call this area the Garbage Patch. One study found that there were six pounds of plastic for every pound of plankton in the Patch. This situation is more than unsightly: It is deadly. Turtles and seabirds, unable to distinguish the plastics from food, end up filling their stomachs with bottle caps, bags, and cigarette lighters. These objects prevent the animals from eating enough to survive, so they starve to death with a bellyful of trash. Plastic also soaks up toxic chemicals like PCBs, poisoning animals from the inside.

Sargasso Sea. They have no way of swimming, so—like the sargassum—they go wherever the wind and currents carry them. The man-of-wars attract a species of fish that feeds on and finds shelter in its tentacles. Young sea turtles also feed on man-of-wars and other jellyfish, and the area around the Sargasso Sea is an important nursery ground for several species of endangered sea turtles. Baby turtles hatch out on beaches around the Atlantic and make their way to the Sargasso Sea, where they may spend several years feeding and growing.

European eels begin their lives in the floating masses of sargassum weed. After hatching, they ride the Gulf Stream north and east, reaching England when they are about three years old. The tiny young eels, known as elvers, migrate upriver in swarms, sometimes digging through wet sandy soil for 30 miles (50 km) to reach upland headwaters and ponds. After a dozen or so years of feeding and growing, adults migrate back to the Sargasso Sea to spawn and lay their eggs. American eels also spawn in the Sargasso Sea, but exit the Gulf Stream earlier than their European counterparts.

CHAPTER 11

LIFE IN THE DEEP

Left: The bottom of Suruga Bay, off Honshu, Japan, 750 feet (roughly 230 m) down. Deep sea organisms make do without sunlight, although some use light from hydrothermal vents for photosynthesis. Top: A bouquet of Corallium. Bottom: Ivory-colored Buccionid whelks. Whelks are carnivores. They build their shells with calcium carbonate extracted from the ocean.

In the deep sea, there is no sunlight. The temperature is near freezing year round, although at hydrothermal vents it may jump to 750°F (400°C) over the space of just a few feet. The pressure, high enough to crush a golf ball, is hundreds of times greater than on the surface of the ocean. Food is scarce. Apart from islands of productivity around vents, seeps, and dead whales, there is virtually no primary production. Small bits of organic matter, known as marine snow, drift down from above. Yet the deep is not entirely featureless and lightless. Enormous seamounts rise up from the abyssal plains, and faults create hills and valleys. Hydrothermal vents emit a dim glow, and a multitude of glowing organisms flash and flicker like stars in the darkness. We know less about deep-sea ecosystems than about any other ecosystem on Earth, and the life we discover here is stranger even than the mythical sea monsters of legend.

The Abyssal Plains

Over a third of the ocean floor is flatter than the state of Kansas. Between continental slopes and mid-ocean ridges, these abyssal plains stretch out for hundreds of thousands of square kilometers at depths averaging between 13,000 and 20,000 feet (4,000–6,000 m). Perhaps because there are so few hard surfaces and so little input of organic material, the density of life here is fairly low. In contrast, the diversity of life is high, and many of the inhabitants are as bizarre as any imaginary aliens from outer space.

Above: A stalked crinoid. Crinoids capture zooplankton using the sticky pinnules covering their crown. They live in both shallow and deep waters and have existed since the Paleozoic era. Top left: A jellyfish in the black depths of the ocean.

MANNA FROM HEAVEN

Organisms living in the deep sea depend on marine snow as a source of nutrition. These clumps of particles that drift slowly down from surface waters include fecal pellets, dust, mucus aggregations, phytoplankton, bits of decaying matter, and radioactive fallout. Microorganisms frequently form colonies on or in the flakes of marine snow. Unfortunately for deep-sea dwellers, microbes and other planktonic organisms consume the vast majority of marine snow before it reaches the bottom. Abyssal animals make efficient use of whatever comes their way, and they likely have slow metabolic rates and fairly low caloric demands.

The most successful group of invertebrates in the deep sea is the echinoderms, at least for animals that can be seen with the naked eye. Brittle stars scoop up particles with their long, elegant arms, while

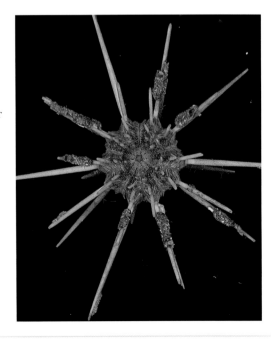

A pencil urchin collected at 1,750 feet (533 m). Tubeworms have fastened themselves to its spines.

wormlike sea cucumbers and sea pigs plow along with their mouth in the muck, although at least one species of sea cucumber has winglike structures that allow it to fly through the water. While some sea urchins stilt-walk on long, elegant spines, others with soft, flexible bodies travel closer to the mud. Sea stars in the deep, like those in the shallows, evert their stomach over whatever they want to digest. Sea lilies on long stalks stretch out feathery arms to feed.

The echinoderms are joined by a multitude of other animals in, on, and above the bottom, including crabs, giant sea spiders, fishes, squid, octopi, jellies, and who knows what else. It seems that every time researchers visit the deep, they find new species. Some, like a recently discovered squid over 6 feet (1.8 meters) long, would be hard to miss in any other ecosystem. Since less than 5 percent of the deep ocean has been explored, a multitude of strange new organisms still remains to be discovered.

LOOKING FOR LOVE

Finding a mate can be difficult, particularly if you live in a perpetually dark and sparsely inhabited region at the bottom of the sea. Abyssal species have a variety of solutions to this problem. Some reproduce asexually, avoiding the need to find a mate in the first place. Others, like urchins, travel in herds, assuring that potential mates are always close at hand. Predatory tunicates, among others, are hermaphrodites,

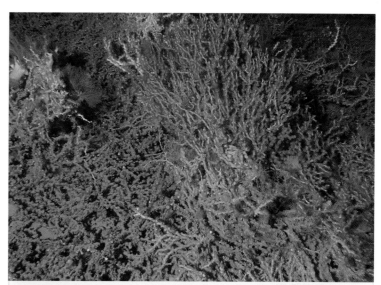

Orange Lophelia, *a coldwater coral that grows in deep-ocean environments.*

DEEP SEA CORALS?

It is commonly believed that corals, because of their reliance on sun and warmth, can survive only in the shallow tropics. However, rich and vibrant coral communities do occur in the cold and dark as well. They have been found as deep as nearly 20,000 feet (6,000 m) and in water as cold as 35.6°F (2°C). While shallow-water corals rely on algal symbionts for a good bit of their energy, their deep-water cousins must get their own nutrition from particulate matter and microscopic life in the water around them. Like tropical reefs, these coral beds are home to myriad other creatures, including crabs, shrimp, anemones, sea stars, and rockfish. Unfortunately, lack of awareness has led to destruction. Fishing boats have dragged heavy dredge and trawl equipment across many coral beds, bulldozing the complex communities. Often, the same trips that document the beauty and diversity of these ecosystems document their destruction by human activity.

producing both sperm and eggs. They can fertilize their own eggs if no mates are around. Deep sea anglerfish form more permanent attachments. Young males latch on to the side of a female, and the couples' tissues eventually fuse together. Most of the male's body degenerates until he is little more than a set of external gonads on the female.

Although sunlight does not make it to the abyss, reproduction may still occur in time with the seasons. In areas where primary production on the surface is strongly seasonal, animal species that produce feeding larvae do time their reproductive cycles to release larvae during the spring peak in marine snow production.

Mountains beneath the Waves

Around the world, tens of thousands of mountains lie hidden beneath the ocean surface. Many of these volcanic peaks, called seamounts, rival the tallest peaks on land, while others are more modest. The tallest mountain on Earth is not Mount Everest, but Hawaii's Mauna Kea, which rises 33,480 feet (10,205 m) from the seafloor. About 800,000 years ago, this mighty peak began as a volcanic eruption on the seafloor. Once it broke the surface of the water, the seamount became an island, but the change may not be permanent. Volcanic islands can sink beneath the waves as they settle deeper into the Earth's crust. Erosion by waves during the sinking process often gives these seamounts flat tops.

Scientists have only the most general notion about the number of seamounts in existence—estimates range from 14,000 to 100,000—and know even less about the organisms that live on them. Studies of seamounts are expanding, however, revealing that they, like many other less-explored areas of the ocean, are home to an astonishing array of organisms.

AN ABUNDANCE OF LIFE

There are a number of reasons why seamounts are so rich

Right: A rock lobster on a reef in the Pacific Ocean. These lobsters are also common seamount inhabitants. Top left: Ely Seamount, in the Gulf of Alaska. A caldera named the "Crater of Doom" is evident at the summit of the seamount.

biologically. First, the bulk of the seamount deflects deep ocean currents up and around it, creating upwelling conditions. This brings extra nutrients into the surface water, increasing primary productivity, which in turn provides more food for animals both in the water above seamounts and on the seamounts themselves. The seamounts may be tall enough that while their bases are in complete darkness, their peaks receive enough light to support photosynthetic organisms. The rocky surface of seamounts provides plenty of hard substrate to which animals like anemones and sea fans may attach themselves. Furthermore,

many sea turtles, fish, and whales seem to use seamounts as navigational waypoints for migration or as meeting spots for feeding and breeding.

Biodiversity varies greatly on seamounts around the world, but it is not uncommon for each seamount to support a number of species that are found nowhere else. On some seamounts, over half the species present are unique to the seamount on which they live. And while it is common for deep-sea vents thousands of miles apart to share the same suite of species, there are two seamount chains off Tasmania that have no species in common, although they are less than 1,900 miles (3,000 km) apart. Some seamount species are "living fossils," or species previously believed to be long extinct. It may be that the relative isolation of seamounts, combined with the availability of ample resources for life, makes them both hot spots for the evolution of new species and refuges

A spoon worm, a sea pen, a stalked crinoid, and two xenophyophores with brittle stars, inhabiting Balanus Seamount in the Atlantic Ocean.

for species that were once more widely distributed in the world's oceans.

THREATENED DIVERSITY

Because they are so rich in commercially important species like orange roughy and rock lobster, seamounts are popular fishing grounds. So, as is the case in so many oceanic habitats, destructive fishing techniques threaten the very ecosystem on which the fish and fisheries depend. One aspect of seamounts that attracts fish is the habitat complexity. Corals and sea lilies can cover as much as 90 percent of the surface. Trawling reduces the percentage of surface covered by these animals to just a fraction of this, and recovery is a long process. Some seamounts in the North Pacific that have not been fished for 50 years still have not recovered. The real tragedy is that hundreds of species unknown to science are disappearing before anybody has a chance to study or even identify them. It is like clear-cutting the rain forests, except even less is known about the diversity being destroyed on seamounts.

Tall, whiplike Lepidisis *bamboo corals with pigtail coils and a rust-colored black coral on Manning Seamount off the coast of New England.*

The Life of a Dead Whale

Craig Smith, professor of oceanography at the University of Hawaii, pays thousands of dollars to sink dead whales to the bottom of the ocean, then thousands more to periodically check in on the rotting corpses. While this might not sound like everyone's idea of fun, Smith has found that these so-called whale falls harbor the densest and most diverse communities yet discovered in the ocean's depths.

What would be a smelly eyesore on the beach provides a bonanza of nutrients in the generally nutrient-poor deep sea. Although whale falls occur unpredictably and probably supply nutrients for no longer than several decades, they may nonetheless be an essential link in the abyssal ecosystem. Whale fall occupants are very closely related to hot vent and cold seep inhabitants, and researchers have suggested that hydrothermal vent animals may use dead whales as alternative habitats. The steep decline in whale populations bemoaned by environmentalists may have repercussions for communities of organisms scientists are just getting to know.

FEEDING FRENZY

Whale-fall communities go through a series of characteristic stages. First, the mobile scavengers like hagfish, sleeper sharks, amphipods, and lithodid crabs (relatives of the king crab) arrive. These animals remove over 90 percent of the whale's soft tissue with remarkable efficiency. In one case, most of the soft tissue from a 30-ton whale was gone within just 18 months. A small whale, on the order of 5 tons, can be cleaned in just four months. While none of these scavengers lives exclusively on dead whales, they certainly do not turn up their noses at such a feast.

When the scavengers are finished, the polychaete worms, snails, and hooded shrimp move in. These opportunist species take advantage of the messy eating habits of the scavengers, feeding on organic matter in the sediments around the carcass as well as on the carcass itself. Some of the animals that appear at this stage have been found only on whale falls, making it seem that they somehow survive by hopping

Below: Whale fall on the seafloor, photographed by Monterey Bay Aquarium researchers in February 2002. The bodies of thousands of Osedax *worms have formed a red "carpet" along the bones. Top left: Protective tubes built by polychaete worms.*

from dead mammal to dead mammal. One group of whale-fall specialists is the bone-eating zombie worms. These worms send rootlike extensions of their bodies into the whale's bones while extending slime-covered, flowerlike plumes into the water. Symbiotic bacteria in the "roots" break down the fatty material that makes up more than half of the bones' weight into a form the worms can use. In some cases, there are so many zombie worms on the carcass that it looks like a shag carpet.

THE FINAL STAGES

During the second stage, bacteria with large, white cells may begin appearing on the bones. These bacteria are chemoautotrophs, microbes that use chemical compounds as their energy source. They eventually form thick mats that are characteristic of the third stage in the life of a dead whale. As is the case at hydrothermal vents, chemoautotrophs, feeding on sulfides and other compounds in the decaying bones, support a rich diversity of creatures. More than 30,000 animals have been found on a single whale carcass during this stage, including mussels, clams, and polychaete worms.

When the organic material is completely used up, the remaining whale bone becomes a reef of sorts, providing hiding places and hard surfaces for animals looking for a home rather than a meal.

Above left: A whale bone blanketed by Osedax frankpressi *worms. The worms' red and white plumes are thought to function as gills. Below left: A beached whale. Whale carcasses are a valuable source of nutrients to deep ocean life, but they are a sad sight and difficult to dispose of when beached.*

Hot Vents, Cold Seeps

In 1977, a trip to the Galápagos Rift zone off the coast of Ecuador yielded one of the most thrilling oceanographic discoveries of the twentieth century: hydrothermal vent communities. Thousands of feet below the surface, shimmering mineral-rich water spewed forth from underwater hot springs, forming a field of dark chimneys surrounded by a rich community of organisms. The showy red plumes of tubeworms around the bases of the chimneys inspired researchers to name this site the Rose Garden. The Rose Garden has since been covered by a layer of lava, but our understanding of deep-sea biology continues to grow.

Above: White flocculent mats, composed of bacteria and diatoms, are visible among the gassy, 212°F (100°C) white smokers at Champagne Vent in the Pacific Ocean. Top left: Mussels, spider crabs, and worms are part of a community that depends on energy from a hydrocarbon seep.

SMOKING HOT

Hydrothermal vents occur where two tectonic plates move away from each other. At these spreading centers, seawater that seeps through cracks in the ocean floor come in contact with magma and hot, newly formed crust. The water dissolves minerals from the rock, heats up, and shoots back through the crust. More than 200 hydrothermal vent fields have been discovered around the globe.

Water is typically around 716°F (380°C) when it comes out of the vent, but temperatures vary considerably. The hottest vent yet discovered produces water that is 765°F (407°C). At this temperature, the water is a supercritical fluid, combining the properties of liquid and gas. The color of vent fluids depends on their chemical composition. Black smokers are rich in iron and sulfide, while the cooler white smokers have more barium, calcium, and silicon. When hot vent water meets cold bottom water, minerals precipitate out of solution to form dramatic chimneys. These chimneys grow quickly, up to 30 feet (9 meters) in 18 months. The largest smoker yet discovered, named Godzilla, reached the height of a fifteen-story building before it collapsed under its own weight.

ALIENS ON EARTH

In contrast to terrestrial food webs, hydrothermal vent communities do not depend on energy from sunlight, and nothing resembling plant life exists in them. Fueling these communities are single-celled chemoautotrophs, microbes that get energy from chemical compounds. Some form dense mats, while others live inside the tissues of clams, mussels, or worms. The symbiosis between host and microbe can be quite intimate: giant tubeworms no longer have mouths or stomachs, depending entirely on the microbes for their nutrition. Although hydrogen sulfide is poisonous to the worms, their symbionts require it, and the worms have a special molecule in their blood that carries hydrogen sulfide from the seawater to the waiting bacteria. Some microbes living beneath hydrothermal vents have set the record for most heat-loving organism, surviving at temperatures up to a whopping 330°F (169°C).

COLD, SLOW, AND STEADY

Not all oases of life in the deep sea are boiling hot. Seven years after the discovery of the first hydrothermal vent, scientists discovered the first cold seep community in Monterey Canyon, off the coast of California. Like hot vents, cold seep communities are powered by microbial digestion of chemical compounds rather than by sunlight. Unlike hot vents, cold seeps are the same temperature as the surrounding seawater, and form where methane, hydrogen sulfide, or oil seeps out of the sediment. In some cases, methane may bubble out fast enough to form dramatic mud "volcanoes," but most seeps are characterized by a slow and steady influx of chemical compounds. Consequently, cold seep animals tend to grow more slowly and live longer than their cousins at hot vents. Indeed, one cold seep worm species may be the longest-lived animal, with some individuals estimated to be over 250 years old.

Photosynthetic red and green algae coexisting with chemosynthetic bacterial mats.

NO SUN? NO PROBLEM.

For years, biologists assumed that since there was no sunlight in the deep sea, there could be no photosynthesis. After all, where could an organism find a strong and consistent source of light beneath several miles of water? Then deep-sea researchers discovered that some hydrothermal vents give off a dim but steady glow known as geothermal light. In 1999, researchers reported that some bacteria at these vents used photosynthesis to supplement the energy they manufactured through chemosynthesis. Still, the question remained whether an organism that required photosynthesis could ever survive in the deep sea. Finally, in 2005, an international team isolated a species of green sulfur bacteria from a hydrothermal vent plume that could only survive in the presence of light and concluded that photosynthesis exists at the bottom of the sea.

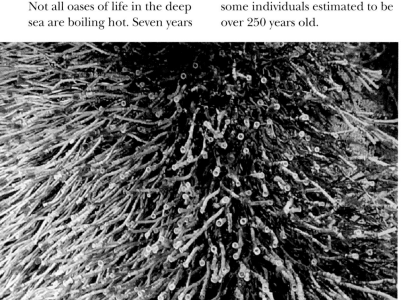

Tubeworms carpeting a lower-temperature sulfide chimney. This chimney is named Zooarium, because it is inhabited by a wealth of undersea life.

Billions of Living Stars

The word "biolumines-cence" means living light. Organisms that are biolumines-cent, or capable of producing light, can be found in many depths and regions of the sea. Scientists estimate that over 90 percent of organisms in the deep are bioluminescent. Organisms ranging from bacteria to squid to fish can bioluminesce, but they do not all do it the same way. Some animals produce light from chemicals stored in their bodies, others glow by harboring sacs of bioluminescent bacteria, and a few, like midshipmen fish, require additional ingredients that they obtain from their diet.

WHY GLOW?

Bioluminescence serves many purposes for deep-sea creatures. In several species of shrimplike ostracods, males attract females by giving off a sequence of flashes as they swim. Like fireflies, their pattern of flashes is species-specific, so females do not waste their time ap-proaching males of the wrong species. Bioluminescence is also important in the predator-prey game. Some deep-sea shrimp eject a glowing cloud when at-tacked, presumably to confuse or repel the predator and give the shrimp time to escape. However, predators use bioluminescence, too. Deep-sea anglerfish dangle a light organ that looks like a tempting food item in front of their heads and make a meal of anyone who comes looking. Some siphonophores, animals that resemble a floating colony of jellyfish, use glowing lures at the ends of clear tentacles to attract food. They jiggle these lures in much the same way that fishermen do. A group of fishes called loosejaws are particularly crafty in their use of bioluminescence. Red light

Above: A physonect siphonophore. Some deep-water siphonophores have dark orange or red digestive systems that can be seen inside their transparent tissues. Many are bioluminescent, glowing green or blue when disturbed. Top left: The glowing eyes and body of the short-nose greeneye fish.

This deep-sea anglerfish was caught near Averoy, Norway.

travels poorly through water and is rare in the deep sea. Most deep-sea animals lack the ability to detect it. Loosejaws, however, both produce red light and are capable of seeing it. By lighting up a red bioluminescent organ near their eyes, they can find prey and communicate with other loosejaws without giving themselves away.

HOW DOES IT WORK?

Unlike lightbulbs, bioluminescence is an extremely efficient way to make light, because it produces virtually no heat. While fluorescence occurs only in the presence of an external light source, and phosphorescence requires an initial input of light energy from an external source, bioluminescence does neither. It also differs from iridescence, which is the physical diffraction of light that produces a rainbowlike effect. So how does bioluminescence work? The basic reaction requires a minimum of three ingredients—oxygen, a type of enzyme called a luciferase, and a type of molecule called a luciferin. The luciferase and oxygen together activate the luciferin, which then releases light. The signal that turns things on may be a physical stimulus such as a fish swimming by, a chemical stimulus like a change in pH, or even the result of very high population density, as is often the case with bacteria. Luciferin and luciferase molecules are not the same in all species. At least five major types of luciferin are known. The specifics of the bioluminescent

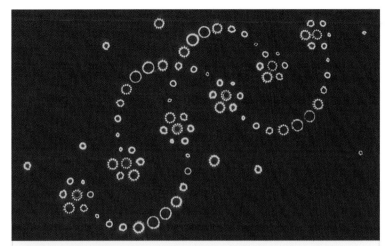

A painting of bioluminescent bacteria, by Angela Bowlds. The uncanny beauty of glowing bacteria has inspired awe in seafarers and scientists alike.

MILKY SEAS

In June of 1854, Captain Kingman of the clipper ship *Shooting Star* observed the so-called Milky Seas phenomenon as he traveled south of Java: "The whole appearance of the ocean was like a plain covered with snow. There was scarce a cloud in the heavens, yet the sky . . . appeared as black as if a storm was raging. The scene was one of awful grandeur; the sea having turned to phosphorus, and the heavens being hung in blackness, and the stars going out, seemed to indicate that all nature was preparing for that last grand conflagration which we are taught to believe is to annihilate this material world."

Although modern observers use less flowery language, Milky Seas continue to inspire awe. They are thought to result from extremely dense concentrations of bioluminescent bacteria. The glow that they generate can cover thousands of square miles, last for days, and be seen from space.

reaction also vary, with many organisms requiring additional compounds.

HUMAN CONNECTION

Bioluminescence is beneficial to humans, too. During World War II, the Japanese used light from ostracods, or seed-shrimp, to help them read maps. Today, doctors and researchers can use bioluminescence to test whether certain bacteria are resistant to antibiotics. They insert a gene for bioluminescence into bacterial cultures. Since some bioluminescence pathways work only if an organism is alive, a culture that continues to glow a few hours after exposure is resistant to the antibiotic used. As scientists continue to study creatures that make light to survive, they gain a deeper understanding of yet another mysterious aspect of life in the oceans.

The Tales Mud Can Tell

The thick layer of sediment at the bottom of the ocean is like an environmental history book. Century after century, material erodes from land, drops from melting icebergs, and falls from the sky. The remains of living organisms become part of the record as well. In the abyssal plains, sediment 3,200 feet (1,000 m) deep can accumulate, while in the trenches the sediment may be several thousand meters thick, dating back hundreds of thousands of years.

One of the tricky parts of studying deep-sea mud is getting samples of it. Remotely operated vehicles or oceanographers in submersibles can pick up small samples. Getting deep cores that go far back in time takes a lot of hollow pipes and a ship with twelve thrusters to keep it precisely positioned while the core is taken. The pipe is lowered toward the seafloor with a two-ton weight attached. When around 25 feet from the bottom, it is allowed to fall, and the weight pushes the corer into the mud. An internal piston pulls up, creating suction to keep the sample in place, and the core is carefully raised back to the ship.

LIFE IN THE MUD

The mud covering much of the ocean floor is not as lifeless as it looks. In just 233 plate-sized samples of sediment taken off the coast of New Jersey and Delaware at a depth of 6,720 feet (2,100 m), biologists found almost 800 species from 14 different animal phyla. Over half of these species were new to science. Because each new sample contained as many new species as previous samples, it is clear that this only scratched the surface of the diversity out there. Based on these results, scientists estimate that one square kilometer of deep sea mud can hold around 10 million animal species.

Animals make up just part of the sediment's diversity. Seafloor mud is also home to single-celled organisms from all three domains of life. Members of at least two of those domains—bacteria and archaea, single-celled life forms without nuclei—may live not just on the surface, but miles down into the oceanic crust. Living microbes have been found in sediment cores more than a third of a mile below the ocean floor, and in hot

Above: The undersea mud slowly claims an abandoned anchor. Top left: Manganese nodules on the floor of the Northeast Atlantic Ocean. Manganese nodules are formed when manganese and iron minerals are deposited on the surface of small undersea objects.

oil reservoirs almost two miles below the North Sea. While chemicals are probably the primary energy source for these organisms, oceanographers from the University of Rhode Island suggest that residual radioactivity may also contribute.

DIGGING INTO PAST CLIMATES

Scientists are using a variety of information from deep sea sediment cores to understand how Earth's climate has changed over the millennia. One approach is to map out how the distribution of cold- and warm-water species shifts over time. Microscopic organisms called foraminifera have been particularly useful in this regard, since species can be identified long after death by the hard shells they leave behind. These shells provide information about ocean chemistry through time as well.

Sediments also provide clues as to how many icebergs there have been over time, which is an indicator of past climates (lots of icebergs means a colder climate). Icebergs begin as glaciers on land, where they accumulate rock, gravel, and other material. When icebergs melt, this material falls to the seafloor and forms characteristic deposits, giving clues to the number and extent of icebergs. In turn, the type of material in these deposits can tell scientists where the icebergs originated. Icebergs that began as Icelandic glaciers, for instance, carry large numbers of dark, glassy, volcanic pebbles, while those from the Gulf of St. Lawrence carry red-stained rocks. Using this type of evidence, scientists have discovered that over the past 40 millennia, huge armadas of icebergs have spread out over the North Atlantic every two or three thousand years, originating simultaneously on both sides of the ocean. This suggests periodic hemisphere-wide cold snaps.

Foraminifera, widespread microbes that live in shells.

NAKED DNA

It turns out that DNA is more than just the vehicle of life's genetic code. It is a source of phosphorous for microbes living in deep-sea sediment. On a global basis, the uppermost 10 centimeters of sediments on the deep-sea floor contain as much as 0.5 gigatons of extracellular DNA, more DNA than is found anywhere else in the world's oceans. Microbes living in the sediment receive about half of their phosphorus by taking apart this DNA and using the phosphorous ions it contains. What is the origin of this naked DNA? Chemical analyses suggest it comes from photosynthetic organisms living near the surface of the ocean.

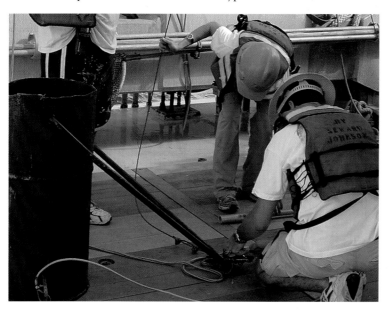

Scientists prepare a pipe dredge, used for extracting samples of undersea sediment. Layers of underwater earth supply crucial information about the natural history of the planet.

CHAPTER 12

THE FUTURE OF OCEANS: THREATS AND SOLUTIONS

In 2003 and 2004, two comprehensive reports on the health of America's oceans were released. Both concluded that ocean ecosystems are in trouble. Some 13,000 beach closings occurred as a result of pollution in 2001. Every eight months, almost 11 million gallons of oil from streets and driveways flows into the nation's seas. Dead zones threaten fisheries and ecosystems in the Gulf of Mexico, the Chesapeake Bay, and other coastal waters. Twenty thousand acres of coastal wetlands disappear annually. Invasive species threaten to drive out natives. Around the world, marine ecosystems face similar threats.

Is this the future of the oceans? Perhaps not. As awareness increases, individuals and organizations step up to the challenge, implementing interesting and creative solutions to these problems.

Left: These oil-soaked eiders came ashore on St. Paul Island in the Pribilofs after the freighter M/V Citrus collided with another ship while loading cargo at sea in 1996. Top: The beautiful but poisonous lionfish was probably introduced to the U.S. east coast by amateur aquarists. Bottom: The Gulf of the Farallones National Marine Sanctuary welcomes visitors and offers educational tours of the intertidal region.

The Global Dumping Ground

According to the United Nations Environment Program (UNEP), 46,000 pieces of plastic litter are floating on every square mile of ocean. Over 400 million gallons of oil run into the ocean annually as a result of improperly discarded motor oil and runoff from roads. Toxic chemicals and fertilizer drain into the ocean through countless rivers and streams. Even radioactive waste ends up in the ocean: 47,000 barrels of radioactive material were dumped near the Farallon Islands, less than 30 miles (48.6 km) from San Francisco.

GETTING AROUND

Pollutants are distributed around the globe by winds and currents, often ending up thousands of miles from their source. Even marine animals can act as carriers of toxic chemicals, contaminating areas where they congregate to reproduce. Researchers found that ponds near large seabird colonies had much higher levels of industrial pollutants like mercury and DDT than ponds farther away, and in some Alaskan lakes, salmon returning to spawn bring high levels of industrial pollutants with them, picked up while feeding at sea thousands of miles away.

UP THE FOOD CHAIN

In 2006, polar bears were named the world's most polluted animals, beating out the previous title-holder, orcas. In part, this stems from the slow degradation of pollutants in the Arctic, where polar bears live, and from the accumulation of chemicals from the industrial Northern Hemisphere that end up in the Arctic via wind and water. However, this is only part of the story. Polar bears have over 70 times the level of some pollutants as their preferred prey, ringed seals. Why is this?

When a pollutant first enters the ocean, a number of things can happen to it. It may break down, accumulate in the sediment, or get carried around by currents. It may also get taken up by living creatures. Microbes, phytoplankton, and protists absorb chemicals from the water directly through their cell walls, and many animals take up chemicals through their gills. While some chemicals are rapidly excreted, others build up over time, becoming much more concentrated in organisms than they are in the water itself. For instance, DDT can be 800 times more concentrated in the bodies of zooplankton than it is in the water. In turn, fish that eat zooplankton

Above: A discarded bicycle becomes a sad part of the underwater landscape. Top left: Polar bears are affected by long-range pollutants transported from industrial areas. These pollutants can lead to infertility.

have around 30 times more DDT in their bodies than the zooplankton, and so on up the food chain. Animals at the top of the food chain may have levels of toxic chemicals over a million times higher than the environment in which they live. This process is known as bioconcentration, and is one reason why the levels of pollution in an environment do not always indicate what the effect will be on organisms within that environment.

PLASTICS

Plastic is everywhere. People use it every day and then throw it away. Every year, millions of pounds of plastic end up in the oceans, where it can stay for a long, long time. Some of it sinks, but some of it forms floating islands of trash. In the middle of the North Pacific Ocean, an area larger than the state of Texas is covered with floating garbage, 90 percent of which is some sort of plastic—everything from Styrofoam to nylon to polypropylene. Many marine animals, particularly birds and sea turtles, die as a result of ingesting plastic. Plastics kill in other ways as well. Close to a million seabirds and 100,000 marine mammals are killed every year when they become tangled in or choke on plastic nets, six-pack beverage holders, or other debris.

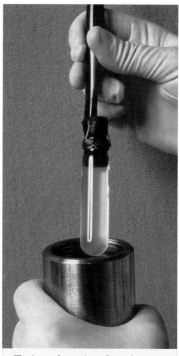

Testing a bacteria culture for use in cleaning up oil spills.

OIL-EATING BACTERIA

Natural seeps release more petroleum into the ocean than spills, runoff, or operational discharge. At a relatively slow and steady rate, day after day, oil seeps out from numerous cracks and fissures in the seafloor. Diverse bacterial communities exist that feed on this oil, and scientists are experimenting with using these oil-eating bacteria to clean up the damaging oil spills caused by humans. This approach avoids the use of chemical dispersants, which can be as toxic as the oil itself. Results so far are mixed, but at least at the shore, adding nutrients to stimulate bacterial growth can speed the breakdown of oil and return the beach to a healthier state.

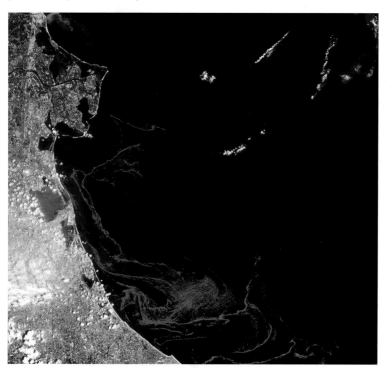

This color-enhanced image of the east coast of Italy shows a large algal bloom in red. The algae feed on the pollutants that pour into the Adriatic Sea from sewage and agricultural runoff.

Habitat Lost

When loggers clear-cut huge swaths of forest, or developers bulldoze grasslands to put in parking lots, the effects are hard to miss. Yet every year, an area of seafloor bigger than Brazil, the Congo, and India combined is essentially bulldozed by dredges and trawlers, and scientists are only just starting to understand what this means for marine ecosystems.

BULLDOZING THE DEEP

Trawling and dredging together account for about a third of all commercially harvested fish and shellfish. Dredging is a bit like raking the bottom. A large metal basket with teeth along its leading edge (the dredge) is dragged along the seafloor to scoop up clams, scallops, and oysters. In contrast, bottom trawling targets mobile animals like shrimp, cod, and flounder. The trawl is a wide, cone-shaped net often equipped with heavy chains or other devices in front to scare up animals that then swim or drift into the net. Both trawls and dredges have devastating effects on seafloor habitat. They crush and knock over most obstacles in their path, including corals, glass sponges, and rock piles that fish and other animals use to hide, breed, and feed. They kick up the sediment, and it has been shown that animals requiring more high-quality organic food become scarce in these conditions. Animals living in burrows in the sediment are rapidly killed. Some areas, like sandy bottoms, may recover relatively quickly from dredging and trawling. The recovery of deep-sea reefs, in contrast, can take centuries, and some structures, like rock piles, never come back. Fortunately, dredging and trawling are not the only options for harvesting bottom-dwellers. Crabs and their kin can be harvested using pots, which do relatively little damage and cause few unintended deaths. Clams can be raked by hand, as is traditional closer to shore. While cutting down trees is a necessary part of logging, the "clear-cutting" of benthic

Above: This turtle excluder device (TED) keeps turtles from getting caught in fishing nets. Top left: Dredging has long been used in harvesting clams. This practice disturbs seafloor communities.

habitats is not a necessary part of harvesting the animals that live there.

KILLING NEMO

Tropical reef fish are an important source of protein for coastal communities throughout the tropics, but they also fetch a high price in the food and aquarium

Two types of apparatus used to catch fish that live near the surface are a purse seiner (top and center) and longliner.

there. While some fish caught this way are sold in the aquarium trade, large numbers also end up in restaurants in Asia, where live fish bring a high price. Unfortunately, the cyanide can poison other reef organisms, potentially causing coral bleaching and death. Dynamite fishing is even more destructive. Dynamite or homemade explosives are thrown into the water, killing or stunning fish and making them easy to gather up. In addition to killing large numbers of animals, dynamite fishing destroys the reef itself, generating huge craters and piles of rubble. Reefs are also dynamited for the rubble itself, which is used for building material and lime, and to create jewelry and curios for tourist shops. Losing reefs means losing some of the world's most breathtakingly beautiful biodiversity hot spots. It also means losing the services they provide, such as food for local communities, protection from erosion and waves, and the potential for ecotourism income.

trades. While people who depend on reefs for food have strong incentives to carefully manage the resource, people who sell fish for a fast profit may use very destructive fishing practices. The two most common are cyanide and dynamite fishing. When cyanide fishing, divers squirt cyanide or other poisons into cracks and crevices to stun the fish that are hiding

What you purchase in the seafood section of your local market has an effect on commercial fishing practices.

WHAT YOU CAN DO

The choices people make when they buy seafood can influence which fishing practices are used and how fisheries are regulated. A number of programs have been developed to help environmentally concerned consumers figure out which seafood to buy. One organization, the Marine Stewardship Council, investigates fishing practices around the world and certifies fisheries and fish suppliers as sustainable. The Marine Aquarium Council provides a similar certification program for fish collected for the aquarium trade. The Seafood Watch program provides wallet-sized cards listing which types of seafood are good, so-so, and bad from a sustainability perspective.

Unintended Victims

About a quarter of what commercial fishermen haul in every year gets discarded as bycatch, species the fisherman had not intended to catch. Some animals are discarded because they are unmarketable or because the fisherman lacks a license for that particular species or the facilities to process it. The longlining method alone kills 40,000 sea turtles and 300,000 seabirds every year, and thousands more become entangled in nets and drown. When it comes to bycatch, the worst offender is probably shrimp trawling. On average,

five pounds of other animals are caught for every pound of shrimp, including herring, crab, flounder, and tuna. In severely overfished areas, the situation is even worse. On shrimp trawlers in the Sea of Cortez, almost 10 pounds (4.5 kg) of other marine life dies for every pound of shrimp captured. Globally, shrimp trawlers generate 19 million pounds of bycatch a year, the equivalent of almost a quarter of the total global commercial fish harvest. Some bycatch is kept and used for fish meal, but most is simply thrown back, dead or dying.

GOOD AND BAD

How many nontarget animals are caught, and what happens to them, depends to a large extent on the fishing method. The most precise technique is harpooning. The harpooner visually identifies each animal before attempting the kill, so nontarget species are rarely if ever harmed. Methods that involve a single hook and line, like trolling or hand-lining, also cause relatively low levels of bycatch. If an unwanted animal is caught, it can be rapidly freed. Large-scale commercial enterprises generally favor less selective but more efficient approaches. Long-liners, typically targeting swordfish or halibut, set out a central line up to 50 miles (80 km) long with thousands of smaller, hooked lines hanging off of it. The hooks are left to "soak" for several hours, so any animal caught on a hook stays there for some time. Birds, turtles, and marine mammals are often caught and drown as a result.

Other fishing methods use nets rather than hooks. A purse seine

Above: A northern sea lion caught in a net that was left behind by a fishing vessel.
Top left: This load of fish was unintentionally caught and killed by a shrimp trawler.

creates a wall of net around a large area that is then closed at the bottom and pulled in. This method is good for catching schooling fish like sardines and tuna, or animals that gather in groups to spawn, like squid. Gill nets are thin, almost invisible curtains of netting placed at various depths in the water. Fish swim into the nets and become ensnared. Sardine, salmon, and cod fishermen often use this approach. Trawlers drag large cone-shaped nets through the water, targeting particular species by altering the depth at which they drag the nets. All of these methods generate large quantities of bycatch. Like long-lining, net-based techniques can trap and drown birds, turtles, and mammals.

GHOST-NETTING

Sometimes, fishermen lose nets or other fishing gear, or discard it intentionally at sea. Whatever the cause, over 600 miles (966 km) of net ends up abandoned in the North Pacific Ocean alone every year. Because nets are made of durable plastic, they can float around for years, catching fish, birds, mammals, and invertebrates. Sometimes nets form large balls that do relatively little damage, but occasionally they continue "ghost-fishing" for years. In the last two decades, almost 200 endangered Hawaiian monk seals were drowned by ghost nets. Net segments over 11,000 feet (3,300 m) long have been found, and segments the length of a football field are not uncommon.

Laysan albatrosses, shown here with a Laysan duck, are sometimes caught and killed by fishing lines when they dive for the baited hooks.

WORKING FOR CHANGE

No one wants bycatch. It threatens populations of marine animals and makes life difficult for fishermen by taking up valuable time on fishing vessels. Working together, fishermen and professors at the University of Washington came up with a relatively simple way to address seabird by-catch in long-line fisheries. Most bird deaths occur when the lines are being released. Birds see the bait on the hooks and attack it, then get dragged down with the lines. It turns out that simply hanging polyester rope with brightly-colored plastic streamers on either side of the long-line as it is deployed scares the birds away. Since 1998, the use of this technique has reduced seabird bycatch by over 80 percent, and the United States Fish and Wildlife Service now makes streamer lines available to long-line fisher-men at no cost.

A sea turtle entangled in a fishing net.

Unlimited No More

There is a long-standing notion that marine animals are "extinction-proof," protected by prolific rates of reproduction, large geographical ranges, and the vastness of the ocean. This view has proved to be false. Twelve marine species are known to be extinct, and 133 have become extinct in parts of their previous range. The Food and Agriculture Organization of the United Nations, the international body charged with monitoring the health of global fisheries, found that a quarter of all fish stocks were overexploited or depleted, meaning that fishing pressure was so high that it could well push some fish to extinction. A further half of stocks are fully exploited, making them dangerously vulnerable to environmental disasters or management errors. The true number of species lost or at risk is doubtless higher, masked by how little is known about even those that are commercially harvested. For instance, scientists realized in 1988 that what was believed to be just two commercially important crab species was really eighteen distinct species, all with unique life histories or ecological niches.

FISHING DOWN THE FOOD CHAIN

The total global fish catch increased steadily until the year 2000, and the numbers still seem fairly robust—down to 90 million tons from a peak of 96 million tons in 2000. Over the

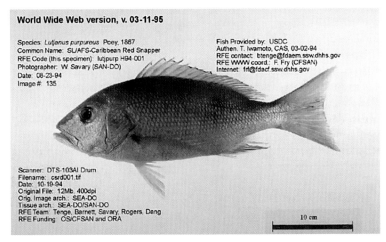

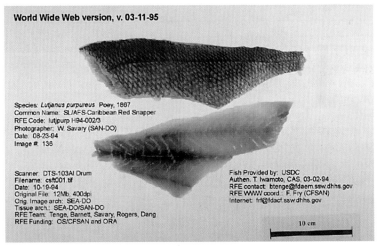

Above: These pages from the FDA's Regulatory Fish Encyclopedia *show the information presented in this resource that helps federal, state, and local officials and consumers to identify species substitution and economic deception in the marketplace. Top left: The popularity of sushi has contributed to the depletion of certain species of fish.*

past several decades, however, there has been a drastic change in the type of fish that make up the global catch. Populations of large, long-lived species near the top of the food web—predators like tuna, cod, and snapper—have been reduced by 90 percent. The fishing industry has compensated by targeting more of the smaller, faster-growing species closer to the base of the food chain.

SMALLER AND SMALLER

Modern agriculture rests on ensuring that individuals with the most desirable traits—large size, for example—are the ones whose genes make it into future generations. Any cattle farmer who killed off the biggest and most desirable cows and let the smaller, less desirable ones produce the next generation would be considered foolish. Yet in many cases, this is just what fishermen are doing. When fishermen harvest a large percentage of the fish in a population using nets, they are removing large fish and letting small fish survive. This has two

effects. First, it instantly reduces the average size of fish in the population. In most fish species, large fish produce significantly more offspring than small ones. For instance, a 23-inch (59 cm) female vermilion rockfish can produce 17 times the number of young that a 14-inch (36 cm) rockfish can. Removing large fish sharply reduces the total number of offspring a population can produce.

Removing large fish can also lead to an evolutionary decrease in fish size. Because big fish are more likely to be killed, individuals with genes for small size and slow growth contribute the most offspring to the next generation. The basic genetic makeup of the population is changed, which decreases the average size of fish for generations. For instance, a fish called tilapia reaches a maximum length of 4 inches (10 cm) and becomes sexually mature at just over an inch in Ghana, where it is actively harvested. Some tilapia were accidentally introduced to Florida, where fishing pressure is much lower. After several generations, the average size of the tilapia had increased to almost 10 inches (25 cm), and fish became sexually mature when they were 6 inches long. Similar effects have been seen in other commercially harvested shellfish and fish, including salmon.

Fishing boats on the Kenai River in Alaska.

Buyer beware. Often, fish labeled as "wild caught" at the grocery store are actually farm-raised.

TRUTH IN ADVERTISING?

Professor Peter Marko, of Clemson University, had heard anecdotal reports that some restaurants were substituting other fish for red snapper, a popular but depleted reef fish. He decided to use this as the basis for a project in a population genetics class he was teaching. Students bought fish fillets labeled as red snapper at nine grocery stores in eight states and analyzed the DNA to see whether the fish had been labeled correctly. The results were stunning. Of the twenty-two fillets purchased by students, just five were really red snapper.

This is not the only case of mislabeled fish. In a recent study, six of eight New York City grocery stores were found to be selling farm-raised salmon as wild-caught. Mislabeling is clearly upsetting to consumers who pay more for "premium" fish. It is also upsetting to conservationists, who argue that it gives a false impression of how well fish stocks are doing. After all, if "red snapper" is readily available at the grocery store, who will believe the species is in trouble?

Too Much of a Good Thing

Ironically, many call it the green revolution—the beginning of the widespread use of chemical fertilizers to boost agricultural productivity. Some argue that without these chemicals, many more people around the world would go hungry. But these same fertilizers, along with waste from hog farms and other sources, are responsible for about 150 so-called "dead zones" in marine waters around the world. In these dead zones, oxygen levels are so low that any animal unable to leave suffocates and dies. Low oxygen levels can also affect the sex ratio of some species of fish, tilting the scales toward more males.

HOW IT HAPPENS

Just as fertilizers boost plant growth on land, they boost the growth of the phytoplankton in the oceans. These so-called algal blooms fuel rapid population growth of the tiny animals that live in the water and feed on the algae. As these animals die, their bodies fall to the bottom, where decomposers set to work. The decomposers use up oxygen. If there are enough of them, they can lower oxygen levels to the point where the animals that normally make their home on the bottom, including crabs, worms, and many species of fish, can no longer survive. Some are able to move to areas

with higher oxygen levels, but the rest die, further fueling the decomposers. Because the growth of algae is fueled by light and warmth as well as nutrients, dead zones are often at their worst in the summer. Although most dead zones are clearly linked to human activities, sediment cores near the mouth of the Mississippi River suggest that four major low-oxygen events occurred there before the use of synthetic fertilizer became widespread. The timing of these events is correlated with periods of high river flow. At these times, a layer of freshwater forms on top of the salt water, and oceanographers suggest that this effectively stops oxygen transport between the atmosphere and the ocean water.

BIG ZONES

The largest dead zone in American waters is in the Gulf of Mexico. The size and timing vary from year to year, but the area of low oxygen can be as large as the state of New Jersey and lasts for weeks or months. This dead zone is linked, in large part, to agricultural fertilizers from the Midwest, carried out to sea via the mighty Mississippi River. The largest dead zone in the world is in the Baltic Sea. It exists year-round and is about four times as

Above: The Baltic Sea contains the world's largest marine dead zone. Top left: A tractor spreading pig manure, which contains nitrogen compounds that can be harmful to the environment if they end up in water sources.

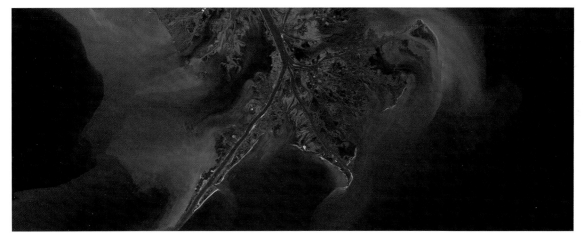

This true-color satellite image shows a dead zone located at the mouth of the Mississippi River as it flows into the Gulf of Mexico.

big as the dead zone in the Gulf of Mexico, covering up to 39,000 square miles (100,000 sq km), an area only slightly smaller than Iceland.

RED TIDES AND SEX CHANGES

Excess fertilizer can also stimulate harmful algal blooms, commonly called red tides. Red tides occur naturally but are more frequent and extensive in areas with heavy fertilizer pollution. They are caused by population explosions of unicellular organisms called dinoflagellates, whose red or brown coloration gives red tides their name. Some dinoflagellates are capable of producing extremely potent neurotoxins that can kill fish, sea turtles, mammals, and birds either directly, or by intensifying in the food chain. The toxins can remain in shellfish and on sea grass long after the bloom is gone. In one case, manatees in Florida died with their mouths full of sea grass that turned out to be coated with red-tide toxin.

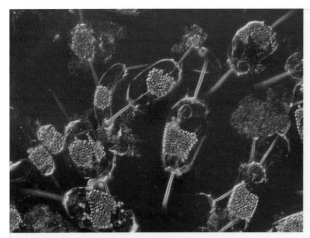

Researchers seeking ways to slow global warming have theorized about the effect of increased iron on phytoplankton growth.

THE GERITOL SOLUTION

In 1989, oceanographer John Martin made a radical claim. He said that in areas of the open ocean with low primary productivity, a good dose of iron would stimulate phytoplankton growth, and that this fact could be used to alter Earth's climate. After all, phytoplankton takes up carbon dioxide, an important greenhouse gas. Grow enough phytoplankton, and presto! A new ice age. Several experiments have shown that adding iron to certain areas of the ocean does indeed produce impressive phytoplankton blooms. Would this work to slow global warming? Probably not. If iron fertilization were implemented on a large scale, the amount of carbon taken up is unlikely to significantly impact global warming and would require putting thousands of pounds of iron into the ocean on an ongoing basis.

Alien Invasions

Tumbleweed is an emblem of the American West, immortalized in song. Waving palm trees in Los Angeles symbolize the Southern California life. Yet these plants are exotics, nonnative species introduced in a variety of ways to an area they did not previously inhabit. All major freshwater, marine, and terrestrial ecosystems in the United States are now home to nonnative species. In a few ecosystems, like San Francisco Bay, exotics now far outnumber natives.

TRANSFORMING ECOSYSTEMS

Not all introduced species take hold in new habitats. Many die, while others survive but do not spread. Of concern to conservation biologists are those exotics that become invasive, rapidly

King crabs—a nonnative species that is thriving in the Barents Sea.

expanding their population, displacing native organisms, and altering habitat from their previous states. Some marine exotics, like the Pacific oyster, are commercially important, but overall they have proved very costly on economic and ecological levels. In a classic tale of disaster, a particularly voracious species of comb jelly named *Mnemiopsis* was accidentally brought to the Black Sea in the 1980s and spread rapidly. By eating most available zooplankton, including fish larvae, it decimated fisheries in the Black, Azov, and Caspian seas. In the Caspian Sea, even top predators like the Caspian seal felt its effects. Fortunately a newer arrival to the region, a different kind of comb jelly, likes nothing better

Above: A sea lamprey wrapped around a lake trout. These aggressive parasites are marine fish that spawn in freshwater. Lampreys, native to Lake Ontario, invaded other Great Lakes in the 1920s. Top left: An invasive species of sea squirt was found on Georges Bank, which lies off the coast of New England, in 2003. The tunicate colony can be seen advancing from left to right over pebble gravel.

than a nice meal of *Mnemiopsis*. Scientists are hopeful that the new invader may keep the old one in check.

Off the coast of New England, an invasive tunicate has blanketed huge swaths of Georges Bank, once an impressively productive area for fishes and scallops. This particular tunicate forms mats of interconnected individuals that choke bivalve beds and make habitats less suitable for bottom fish. It has no known predators, and it is unclear what, if anything, can keep it in check.

HOW THEY GET HERE

Some species are intentionally introduced to new regions. After Washington State's native oyster population was severely reduced by overharvesting, for instance, the American oyster was introduced from the east coast. When that species failed to thrive, the Pacific oyster was brought over from Japan. This species is now the mainstay of the regional aquaculture industry and has taken over most of the good intertidal oyster habitat in the region. Intentionally introduced species are often accompanied by unwanted hitchhikers. When Pacific oysters were brought to Washington State, a predatory snail and a parasitic flatworm came along, too. The Japanese oyster drill now preys on oysters as far south as California, and the flatworm has become a major problem for oyster culture in parts of Washington State.

Many species, including barnacles, sea stars, and tunicates, as well as countless other organisms, travel the world on the hulls of boats. If they drop off or reproduce when boats are docked or moored, they may successfully colonize new areas. A far bigger source of invasive species, however, is ballast water. To provide stability, ships traveling empty often fill huge tanks with seawater, which is full of planktonic organisms, including embryos and larvae of benthic animals. When the ship arrives in a new harbor and takes on cargo, ballast water and all the creatures it contains are dumped into the harbor. An average of two million gallons of ballast water an hour is dumped in U.S. waters alone. More than 100 species have been introduced this way, including organisms causing harmful red tides. A 1991 cholera epidemic in South America was linked to the release of contaminated ballast water.

The Japanese oyster drill is a small snail that was imported accidentally from Japan to the California coast early in the twentieth century. These marine snails eat oysters by drilling small holes through the shells of young oysters with a ribbonlike band of teeth.

Climate Change

There is no longer much doubt that Earth is warmer than it has been at any time in the past thousand years, and perhaps at any time in the past 650,000 years. While some of the warming may be natural, a significant portion is attributable to human activity, primarily the use of fossil fuels. Implications for marine ecosystems are wide-ranging and severe. Species are appearing at higher latitudes than normal. Upwelling events that sustain rich coastal fisheries are occurring less predictably. Meeting the challenge of climate change will require a concerted global effort.

RISING TIDE

As glaciers and ice caps melt, water previously stored on land ends up in the ocean, raising sea levels. If the entire Greenland ice sheet melted, that alone would increase sea level by 22 feet (6.5 m). The East Antarctic ice sheet could increase sea level by a whopping 213 feet (65 m). Sea level is also increasing due to thermal expansion of seawater. The rate of sea-level rise has doubled in the last century and a half, and models suggest that by 2100 the average sea level will be more than a foot higher than it was in 1990. Along gently sloping shorelines, this would result

in a shoreline retreat of almost 1,000 feet (300 m).

Rising seas displace people and ecosystems. Four islands in India's Sundarbans have already disappeared beneath the waves, and residents of some low-lying Pacific Island nations are contemplating a mass migration to Australia or New Zealand. The fossil record documents widespread loss of mangrove forests during times of rapid sea-level rise, a situation that may soon be repeated. Past increases in sea level changed the California coast from productive rocky reefs to the sandy shores that exist today.

DISAPPEARING SEA ICE

Like glaciers and ice caps, Arctic sea ice is disappearing rapidly. It forms later in fall, melts earlier in spring, and covers a smaller and smaller area each year. Indeed, the Arctic Ocean may be ice-free by the end of the century. While this could help commerce by opening a trans-Arctic shipping passage, it will devastate people and other animals that rely on sea ice. The absence of sea ice during winter storms has led to dramatic increases in coastal erosion in the Arctic, forcing some villages to abandon

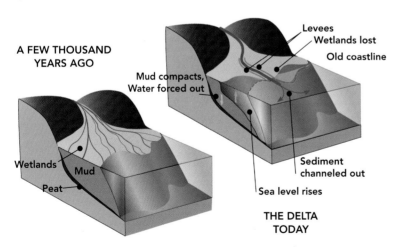

Above: In the Mississippi Delta, many forces combine to create a higher-than-average rate of sea level rise. Channelization of the river stops the influx of new sediment, while older sediment compacts and sinks under its own weight. Top left: An Adelie penguin walks on ice in the Ross Sea.

A ringed seal rests on the ice.

sites where they have existed for thousands of years. Because polar bears hunt from sea ice, the decline in ice cover is leading to starvation for many, and scientists have seen drowning polar bears for the first time. Ringed seals give birth and raise their young on sea ice, and decreases in sea ice and snowfall have led to lower survival rates of seal pups in many areas of the Arctic.

This area near Maryland's Patuxent River flooded as a result of extremely high tides. If sea level continues to rise, this could become a common occurrence.

Higher sea-surface temperatures have caused the bleaching of many shallow-water corals, like this table coral in Papua New Guinea.

DYING REEFS

Tropical corals get their color as well as most of their energy from microscopic algae inside their cells. When corals are stressed by too much light, heat, or disease, they expel the algae in a process called bleaching. Corals can recover from bleaching under some circumstances, but they frequently die. As tropical ocean temperatures have increased in recent decades, scientists have seen an increase in mass coral bleaching events. In 1997 and 1998, widespread bleaching occurred in reefs around the world. Some areas, such as Sri Lanka, Singapore, and parts of Tanzania, lost over 90 percent of their shallow-water corals. Caribbean reefs suffered severe mortality in 2005 and 2006, even among species that are normally resistant to bleaching. Over 95 percent of lettuce coral, 93 percent of star coral, and 61 percent of brain coral were bleached, and some coral colonies over 800 years old did not survive. The future of shallow-water reefs looks grim indeed, prompting biologists to consider actions as extreme as covering reefs with shades during particularly hot periods.

ACID BATH

A particularly insidious effect of climate change is a gradual increase in the ocean's acidity resulting from increased carbon dioxide. Just as acid dissolves bones and teeth, a lower, more acidic pH slows the rate at which organisms can build calcium carbonate structures like shells and the massive, reef-forming coral skeletons. Because cold water can hold more carbon dioxide, acidification will be most severe in polar waters, though it could have long-lasting effects throughout the ocean. By 2050, the pH of tropical waters could drop enough to slow coral growth rates by half. By 2100, polar waters may be acidic enough to dissolve the shells of planktonic snails.

Protecting the Commons

In the early 1600s, Dutch jurist and philosopher Hugo de Groot formally laid out the doctrine known as "freedom of the seas" in response to conflicts between the Dutch East India Company and Spain, Portugal, and England. Coastal nations had sovereignty over the area within a cannon shot of their shores, around 3 nautical miles (6 km), but beyond that it was open territory. A period of unilateral expansion of coastal sovereignty by a number of nations in the middle of the twentieth century prompted renewed international discussion on the issue. The United Nations Convention on the Law of the Sea (UNCLOS), an international treaty, finally hammered out in 1982, came into effect in 1994. UNCLOS defined an exclusive economic zone (EEZ) extending 200 nautical miles (370 km) from shore in which a nation has sovereign rights to all natural resources, and laid out an international obligation to safeguard the marine environment. Enforcement and agreement are relatively difficult to achieve on a global level. A few international treaties have gone into effect that will help protect the world's oceans—one to reduce the occurrence of persistent organic pollutants like DDT and PCBs, another to limit species introductions through ballast water—but much work remains to be done.

NATIONAL EFFORTS

Given the power both to exploit and protect marine resources within their EEZ, nations have done a mix of both. Resource management tends to focus on individual species, setting catch or equipment limits or banning harvest of a particular species altogether. The success of this approach depends on sufficient understanding of the ecology and current status of the species in question, and on the willingness of fishery officials to set and enforce catch limits based on science, despite immense pressure from the fishing industry. While this approach has worked for some commercially harvested populations, other populations have suffered devastating declines due to inappropriate management.

PROTECTED AREAS

A broader approach to management and conservation focuses on places rather than individual species. After all, species cannot survive in the wild without appropriate habitat or the biological community of which they are

Above: The Channel Islands National Marine Sanctuary, located off the coast of California. Top left: One of the first doctrines to regulate ocean territories was set out in the early 1600s, around the time that Henry Hudson's Half Moon *arrived in the Hudson River.*

a part. Less than 1 percent of the world's ocean is currently designated as a protected area, but this is rapidly changing. In June 2006, 137,792 square miles (222,000 sq km) of the Pacific Ocean around the northwest Hawaiian Islands was declared a marine reserve, making it the largest in the world. That same summer, the state of California put in place a net-

Tomales Point, at the Gulf of the Farallones National Marine Sanctuary. The sanctuary covers 948 square nautical miles (1,255 square miles) off the northern and central California coast.

work of marine reserves that had been six years in the planning, banning or restricting fishing in almost a fifth of the state's marine water. Bottom trawling has been outlawed in more than 60 percent of the Aleutian Islands' fishing grounds.

Marine reserves have many potential benefits. By providing fish and other animals a safe haven for growth and reproduc-tion, they can increase the size and number of fish both in the reserve and in adjacent waters. Five years after five small reserves had been created around St. Lucia, the number of fish near the reserves had increased by 50 to 90 percent. While the number of marine reserves is increasing, conservationists worry that many are so-called paper parks—parks that exist on paper but lack sufficient monitoring and enforcement. Some supposedly protected areas even allow damaging resource use within their boundaries. While almost a fifth of the world's coral reefs have been designated as reserves, a survey of managers and scientists concluded that just 1 in a 1,000 reefs can be considered truly protected.

The Alaska halibut fishery has successfully implemented a limited-access approach to managing the industry's resources.

PROBLEM OR PANACEA?

One hotly debated approach to fisheries management is shifting from a centrally managed open-access approach to limited access property rights. In theory, having ownership of the resource gives fishermen incentive to manage it sustainably. In the Tasmanian rock lobster fishery, the number and size of individuals as well as egg production increased dramatically following the institution of rights-based management. In the Alaska halibut fishery, conditions are safer, bycatch is lower, and fish populations are higher compared with earlier management approaches. Success is not universal, though, and many worry that improperly administered rights-based programs will consolidate what should be a public trust (fisheries) in the hands of a few large corporations.

Hope for the Future

A growing awareness of the troubled state of the world's oceans has inspired action. The challenges are formidable, no question, but the situation is not hopeless. With humor, determination, and creativity, individuals, communities, and governments around the world are making a difference. Both understanding and conservation of the sea and its inhabitants are increasing daily.

NEW DISCOVERIES

While some turn their eyes to the skies in search of the unknown, there is no shortage of mystery in the ocean. Species new to science, some quite bizarre, are discovered regularly. An entire coral reef was found off Thailand in 2006, reminding scientists how poorly mapped even the shallow seas are. In the deep sea, complex communities of single-celled organisms have been discovered living deep in the rock of the oceanic crust. Canadian and American researchers are planning an ambitious undersea observatory that may eventually cover thousands of square miles of seafloor off the Pacific coast of North America. The first stage, a 500-mile (800 km) ring of fiber optic cable and two remotely controlled laboratories, is already under way. When complete, this program promises to revolutionize the study of biological, chemical, physical, and geological oceanography.

Above: A scuba diver, tethered to other divers for safety, collects samples. Top left: This gray whale was one of three found trapped in the ice in the Bering Sea. A joint effort by Americans and Russians ultimately saved two of them.

The blue goose is the mascot of International Migratory Bird Day, a conservation education event in Fairbanks, Alaska.

INTERNATIONAL EFFORTS

The resources needed to understand and protect the world's oceans are beyond the means of any one country, leading to the emergence of a number of strong international collaborations. One focus has been migratory species, whose travels regularly take them across international boundaries. Southern resident killer whales, for instance, spend their days in both British Columbia and Washington. These whales have recently been listed as endangered by both Canada and the United States, and people from both sides of the border are working together to save them. The largest remaining Pacific populations of the critically endangered leatherback turtle inhabit the beaches of Papua New Guinea, Indonesia, and the Solomon Islands, and governments of these three countries have pledged to share their knowledge and resources in an effort to stave off extinction of this species.

Many ecosystems likewise span multiple countries, and multinational collaborations are emerging to protect large swaths of sea. Malaysia, the Philippines, and Indonesia are building a joint agreement to protect the Sulu-Sulawesi Seas, a biodiversity hot spot ringed by the three countries. The European Union has banned trawling around several deep-sea reefs near the Madeira, Azores, and Canary Islands, and Mediterranean nations have agreed to stop bottom trawling in sensitive areas off Italy, Cyprus, and Egypt.

RECOVERING SPECIES

While many marine species are in dire straits, some have experienced spectacular recoveries. The eastern gray whale, once hunted to near extinction, was removed from the endangered species list in 1994. Overhunting reduced northern elephant seals to just a handful of individuals at the turn of the nineteenth century, but today there are more than 30,000 of them. Bowhead whale and California sea lion populations are likewise increasing. Striped bass, once teetering on the edge of doom, were declared fully recovered in 1995, and lower catch limits and closed nursery areas brought a rapid recovery of North Atlantic swordfish populations. Not all endangered species can be saved, but with effective conservation plans and adequate enforcement, many can.

Steller's sea lions are endangered, with current population declines likely related to overharvesting of their preferred prey by commercial fishing fleets.

GLOSSARY

A

ABYSSAL Pertaining to the ocean depths between approximately 13,000 and 16,500 feet (4,000 and 5,000 m).

AEROBIC Requiring or using oxygen.

ALGAE Collective term for marine and freshwater organisms with chlorophyll and capable of photosynthesis, including seaweed (singular: alga).

ANAEROBIC Occurring in the absence of oxygen.

ANOXIC Without oxygen.

APHOTIC ZONE Area of the ocean below approximately 600 feet (200 m), too deep for any light to penetrate.

ARCHAEA Most recently recognized of the three domains of life; single-celled organisms without nuclei, often found in extreme conditions. Previously grouped with bacteria, but are no more closely related to bacteria than to eukaryotes.

ARTHROPODA The largest animal phylum, whose members are characterized by a hard external skeleton and jointed appendages; includes horseshoe crabs, lobsters, and krill.

ASTHENOSPHERE The hot, deformable layer just below Earth's crust; the uppermost layer of Earth's mantle.

AUTOTROPH An organism that makes its own food through photo- or chemosynthesis; an organism that gets its carbon from carbon dioxide.

B

BACTERIA One of the three domains of life. Single-celled organisms without nuclei.

BATHYAL Pertaining to the ocean depths between approximately 700 and 13,000 feet (1,000–4,000 m).

BENTHIC Of or pertaining to the seafloor.

BIODIVERSITY Biological diversity; may be measured on a variety of levels, including number of species, number of phyla, or genetic diversity.

BIOLUMINESCENCE Light produced by living creatures.

BIOSPHERE All the areas of Earth's crust, water, and atmosphere where life exists.

BIVALVE Collective term for mollusks having paired shells, including clams, mussels, and oysters.

BYCATCH Organisms unintentionally caught along with the species targeted for harvest.

C

CHEMOAUTOTROPH An organism that makes its own food through chemosynthesis; an organism that gets energy from inorganic matter and carbon from carbon dioxide.

CHEMOSYNTHESIS The creation of organic material using inorganic material as an energy source.

CHLOROPHYLL A pigment, usually green, used to capture light energy for photosynthesis.

CLASS The unit of biological classification one level below phylum and one level above order.

COLD SEEP Seafloor habitat where organic materials like methane and hydrogen sulfide seep up from below, often fueling rich biological communities.

COMMENSALISM A type of symbiosis where one organism benefits and the other neither benefits nor is harmed.

COMMUNITY A group of organisms of several species that live in the same area and interact regularly.

CONSUMER Organism that gets its energy by eating others.

CONTINENTAL MARGIN Underwater extension of a continent; made up of the continental shelf and continental slope.

CONTINENTAL RISE A gently sloping wedge of sediment at the transition between the continental slope and the abyssal plains.

CONTINENTAL SHELF Relatively shallow, gently sloping extension of a continent into the ocean just above the continental slope.

CONTINENTAL SLOPE The true edge of a continent; sloping region between the relatively shallow continental shelf and the abyssal plains.

CONVERGENCE ZONE Places in the ocean where two different water masses meet.

COPEPODS Small, planktonic crustaceans; important component of most shallow-water food webs.

CORIOLIS EFFECT The apparent deflection of moving objects from their original path due to Earth's rotation; currents are deflected to the right in the Northern Hemisphere and to the left in the Southern Hemisphere.

CRUSTACEAN Member of the phylum Arthropoda, including lobsters, crabs, shrimp, and copepods.

D

DECOMPOSER Organism that feeds on and breaks down the dead bodies of other organisms.

DEEP SCATTERING LAYER A dense aggregation of fish and other animals deep in the ocean that reflects sonar, creating a false bottom.

DETRITIVORE Organism that eats detritus.

DETRITUS An accumulation of decaying organic matter.

DIATOM The most abundant type of phytoplankton; single-celled photosynthetic organisms with two silica valves.

DINOFLAGELLATE One of the most abundant types of phytoplankton; a single-celled organism with a pair of flagella may be capable of both photosynthesis and heterotrophy; cause of red tides.

DOMAIN The highest level of biological classification. Each of the three domains (Archaea, Bacteria, and Eukaryota) contains many kingdoms.

E

ECHO SOUNDER A device that senses water depth by bouncing sound waves off the bottom.

ECOSYSTEM Community of organisms and their nonliving environment.

ESTUARY A partially enclosed body of water where rivers meet the sea.

EUKARYOTA One of the three domains of life; made up of all organisms with a nucleus, including plants and animals.

EUTROPHICATION The physical, chemical, and biological changes that occur when water receives too many nutrients.

EVOLUTION Change in the frequency of inherited characteristics within a population of organisms over several generations; may be adaptive or neutral.

F

FAMILY A unit of biological classification one level below order and one level above genus.

FILTER FEEDER *See suspension feeder.*

FOOD CHAIN *See food web.*

FOOD WEB A complex set of associations between organisms that eat and are eaten by each other; the flow of food energy through an ecosystem.

FORAMINIFERA Group of amoeba-like single-celled organisms often encased in a calcium carbonate shell; fossil foraminifera are important indicators of past climates.

FOSSIL FUEL Fuel derived from long-dead organisms; includes oil, coal, and natural gas.

FREE WAVE A progressive wave no longer acted upon by the forces that formed it.

G

GAMETE Sex cell; sperm or egg.

GASTROPODA A class within the phylum Mollusca whose members are characterized by a single muscular foot and a distinct head with sensory organs; includes snails and sea slugs.

GENUS Unit of biological classification one level below family and one level above species.

GUYOT A flat-topped submarine mountain; formed by the erosion of an island as it sinks beneath the waves.

GYRE A circular flow of water around an ocean basin; five major gyres exist in the oceans.

H

HADAL Pertaining to the ocean depths below approximately 19,800 feet (6,000 m).

HERMAPHRODITE An organism producing both male and female sex cells.

HETEROTROPH An organism that, unable to synthesize its own food, gets its energy from other organisms.

HYDROTHERMAL VENT A deep-sea hot spring; an area in the deep sea where hot, mineral-rich water flows up through the oceanic crust; often supports a rich and bizarre community of organisms.

I

INTERNAL WAVE A wave that forms underwater between layers of different density.

INTERTIDAL Between high- and low-tide levels; regularly submerged and exposed by the rise and fall of the tides.

K

KINGDOM An order of biological classification second only to domains; eukaryotes have traditionally been divided into four kingdoms.

L

LITHOSPHERE The rigid, relatively cool outer layer of Earth; includes oceanic and continental crust as well as the upper layer of the mantle.

LITTORAL *See intertidal.*

LONGSHORE TRANSPORT The wave-driven movement of sediment parallel to the shore.

M

MACROALGAE Seaweed; multicellular algae visible to the naked eye.

MANGROVE A diverse group of trees that characterize tropical coastlines; provide protection against erosion, storms, and tsunamis.

MANTLE The layer of Earth between the crust and the core.

MARINE SNOW Clumps of organic material that fall through the water column; a valuable source of nutrition for deep-sea organisms.

MICROBE A microscopic organism.

MID-OCEAN RIDGE A relatively shallow area in ocean basins formed where two tectonic plates are moving apart from each other.

MOLECULE A chemically bonded group of atoms; the smallest unit of a substance that still maintains the basic characteristics of that substance.

MOLLUSCA A phylum of animals that includes snails, sea slugs, clams, and mussels.

MULTIBEAM SONAR *See sonar.*

MUTUALISM A symbiosis in which both partners benefit.

N

NATURAL SELECTION Mechanism of adaptive evolution caused by the differential reproductive success of individuals within a population.

NEKTON Animals that live in the ocean and are relatively strong swimmers; includes fishes and whales.

NERITIC Of or pertaining to the shore or coast; used to refer to continental margins, the overlying water, and the organisms that live there.

NONTARGET SPECIES Unwanted fishes and animals caught accidentally in fishing gear.

O

OMNIVORE An animal that eats both plants and animals.

OPPORTUNIST An animal that takes advantage of whatever food sources are available.

ORDER A unit of biological classification one level below class and one level above family.

ORGANIC COMPOUND A compound that contains carbon and at least two other elements, often hydrogen, nitrogen, or oxygen.

ORGANISM Any living thing, from the smallest bacterium to trees to whales.

P

PARASITISM A symbiosis in which one partner benefits to the detriment of the other.

PELAGIC Pertaining to the open ocean.

PHOTIC ZONE The upper layers of the ocean where light can penetrate; approximately from the surface to 600 feet (200 m).

PHOTOAUTOTROPH An organism capable of producing organic compounds using light energy; forms the base of many food webs.

PHOTOSYNTHESIS The creation of organic material using light as an energy source.

PHYLUM A level of biological classification just below kingdom and above class.

PHYTOPLANKTON Planktonic organisms capable of photosynthesis; includes algae and bacteria.

PLANKTON Drifting or weakly swimming organisms that live in the water column.

PLATE BOUNDARY The boundary between two tectonic plates.

PLATE TECTONICS The theory that the surface of Earth is made up of several distinct plates whose movements are driven by convection currents within the mantle.

POPULATION A group of individuals of the same species who live in the same area and interact regularly.

PRIMARY PRODUCER An organism capable of producing organic compounds using light or chemical energy; forms the base of the food web.

PRIMARY PRODUCTIVITY The amount of organic material produced by chemo- or photosynthesis in a given area; determines the ability of an area to support animals.

PROKARYOTE An organism without a nucleus; includes bacteria and archaea, two distinct domains.

PROTIST Member of the kingdom Protista within the domain Eukaryota; a eukaryote that isn't a plant, animal, or fungus.

PYCNOCLINE An ocean zone where water density changes significantly over a relatively short distance.

S

SCAVENGER An animal that eats creatures that have already died.

SEAGRASS A true (vascular) plant that lives in shallow seas; has roots, flowers, and seeds.

SEAMOUNT A submarine mountain.

SEICHE A standing wave occurring in an enclosed or semi-enclosed body of water; like water sloshing in a bathtub.

SESSILE Attached to the substrate.

SONAR Sound navigation and ranging; equipment that uses sound to visualize solid objects underwater; may emit single or multiple "pings."

SPECIES The basic unit of biological classification; a group of closely related organisms. Often defined as "a group of actually or potentially interbreeding individuals," but this definition applies primarily to sexually reproducing animals.

SPLASH ZONE The shore zone just above the intertidal, wetted only by rain and splash from the waves.

STANDING WAVE A wave in which water goes up and down without seeming to move forward.

SUBDUCTION ZONE A region where one tectonic plate goes beneath another.

SUBSTRATE A surface on which a sessile organism might grow; may be rock, wood, metal, etc.

SUPRALITTORAL *See splash zone.*

SUSPENSION FEEDER An animal that feeds on particles suspended in the water.

SWASH The wave-driven movement of water up a beach.

SYMBIOSIS A relationship wherein organisms of different species live together in a close, long-lasting association; may be a mutualism, parasitism, or commensalism.

T

TEMPERATE ZONE The middle latitudes; the region of Earth between the tropical and the polar regions.

TERRESTRIAL Of or pertaining to the land.

TIDAL BORE A large wave formed in areas where the tidal crest moves rapidly up a river or estuary.

TIDAL RANGE The difference in height between high and low tide.

TRANSFORM FAULT A fault created where two tectonic plates slide past each other.

TURBIDITY CURRENT Rapid underwater avalanche of dense, sediment-laden water; thought to create submarine canyons.

U

UPWELLING Pattern of water movement bringing deep, cold, and usually nutrient-rich water to the surface.

W

WAVELENGTH Distance between two successive wave crests.

WAVE PERIOD Time it takes for two successive wave crests to pass the same point.

Z

ZOOPLANKTON Animals that live in the ocean and drift or swim weakly.

FURTHER READING

BOOKS AND REPORTS

Anderson, Genny. *Welcome to Marine Science.* Online textbook: www.biosbcc.net/ocean/marinesci/mstoc.htm

Broad, William. *The Universe Below: Discovering the Secrets of the Deep Sea.* New York: Touchstone, 1998.

Carson, Rachel. *The Sea Around Us.* New York: Oxford University Press, 1989.

Garrison, Tom. *Oceanography: An Invitation to Marine Science.* Belmont, CA: Thomson Brooks/Cole, 2005.

Kious, W. Jacquelyne, and Robert Tilling. *This Dynamic Earth: The Story of Plate Tectonics.* Washington, DC: US Government Printing Office, 1996. Online edition: pubs.usgs.gov/gip/dynamic/dynamic.html#anchor10790904

Kunzig, Robert. *Mapping the Deep: The Extraordinary Story of Ocean Science.* New York: W. W. Norton, 2000.

Pew Oceans Commission. *America's Living Oceans: Charting a Course for Sea Change.* 2003. Available online through www.pewoceans.org

Safina, Carl. *Song for the Blue Ocean.* New York: Henry Holt and Co., 1998.

Soule, Michael, Elliot Norse, and Larry Crowder. *Marine Conservation Biology: The Science of Maintaining the Sea's Biodiversity.* Chicago: Island Press, 2005.

Steinbeck, John. *Log from the Sea of Cortez.* New York: Penguin Twentieth Century Classics, 1995.

Stewart, Robert H. *Introduction to Physical Oceanography.* Open Source Textbook: oceanworld.tamu.edu/resources/ocng_textbook/contents.html

Sverdrup, Keith, Alan Duxbury, and Alison Duxbury. *An Introduction to the World's Oceans.* Boston: McGraw Hill, 2005.

Tomczak, Matthias. *Shelf and Coastal Oceanography.* 1998. Online textbook://gyre.umeoce.maine.edu/physicalocean/Tomczak/ShelfCoast/index.html

United States Commission on Ocean Policy. *An Ocean Blueprint for the 21st Century.* 2004. Online edition: www.oceancommission.gov

WEB SITES

Beachcomber's Alert
www.beachcombers.org
Fascinating information on all sorts of things that have been found floating in the ocean

The Bioluminescence Web Page
www.lifesci.ucsb.edu/~biolum
Fabulous photographs of glowing creatures

Into the Abyss
www.pbs.org/wgbh/nova/abyss
Companion site for a NOVA program on deep sea exploration; excellent descriptions of an oceanographic research cruise

MarineBio.org
marinebio.org
Provides a wealth of information about marine organisms, ecology, and conservation

Marine Biology Web
life.bio.sunysb.edu/marinebio/mbweb.html
A good resource for finding internships, field stations, college programs, and more

Monterey Bay Aquarium Exhibits
www.mbayaq.org/efc
Information and photographs about ocean ecosystems organized by habitat type

NOAA Paleoclimatology
www.ncdc.noaa.gov/paleo/paleo.html
Introduction to techniques and applications of paleoclimatology; includes links to data

Ocean Explorer
oceanexplorer.noaa.gov
An excellent site run by the National Oceanic and Atmospheric Administration, with videos and information on a host of ocean-related topics

Seafood Watch Program
www.mbayaq.org/cr/seafoodwatch.asp
Information about how seafood is harvested and
which types of seafood are environmentally sound
choices to eat

Shifting Baselines
sbflixcontest.org/indexHome.php
A humorous but educational site about marine
conservation

Tides and Currents
tidesandcurrents.noaa.gov/index.shtml
The education section provides several good overviews
of tides and tidal currents

Tree of Life
tolweb.org/tree/phylogeny.html
The latest information on the diversity of life and how
organisms are related

United Nations Atlas of the Oceans
www.oceansatlas.org
Information about the oceans (resource use, geography,
issues, background); geared for policy makers

University of California/Berkeley Museum of
Paleontology
www.ucmp.berkeley.edu
Information about evolution, Earth history, and
geologic time

ORGANIZATIONS AND INSTITUTIONS

Australian Institute of Marine Science
(Townsville, Queensland, Australia)
www.aims.gov.au

Census of Marine Life
www.coml.org

Friday Harbor Laboratories (Friday Harbor, San Juan
Island, Washington, United States)
depts.washington.edu/fhl

Lamont-Doherty Earth Observatory
The Earth Institute at Columbia University
www.ldeo.columbia.edu

Marine Aquarium Council
www.aquariumcouncil.org

Marine Biological Laboratory (Woods Hole,
Massachusetts, United States)
www.mbl.edu

Marine Conservation Biology Institute
www.mcbi.org

Marine Stewardship Council
www.msc.org

Monterey Bay Aquarium Research Institute (Moss
Landing, California, United States)
www.mbari.org

The Ocean Conservancy
www.oceanconservancy.org

Plymouth Marine Laboratory (Plymouth, England)
www.pml.ac.uk

Scripps Institution of Oceanography
sio.ucsd.edu

SeaWeb
www.seaweb.org/home.php

Surfrider Foundation
www.surfrider.org

United States National Estuary Program
www.epa.gov/nep/about1.htm

Wildlife Conservation Society
www.wcs.org

Woods Hole Oceanographic Institution
(Woods Hole, Massachusetts, United States)
www.whoi.edu

WWF International (formerly known as the
World Wildlife Fund)
www.panda.org

AT THE SMITHSONIAN

While best known for its museums, the Smithsonian Institution runs an impressive network of marine laboratories and long-term research sites from Maryland to Panama. These facilities provide a wealth of opportunities for researchers and students interested in the coastal zone and its inhabitants, with a particular focus on biodiversity, evolution, ecosystem dynamics, and environmental change. The Smithsonian is also interested in promoting awareness and conservation of marine ecosystems through outreach, museum exhibits, symposia, films, and books like this one.

SMITHSONIAN MARINE STATION AT FORT PIERCE

www.sms.si.edu

Located on the edge of the Indian River Lagoon on Florida's east coast, the Smithsonian Marine Station sits in a transitional zone between temperate and tropical ecosystems. The lagoon is home to over three thousand species of plants and animals, more than any other estuary in the nation, and was designated an "estuary of national significance" by the U.S. Environmental Protection Agency in 1990. At the marine station, more than one hundred scientists and students visit each year from around the world to study the diversity of the lagoon and the physical processes that support this diversity. The dedication of researchers can be amazing: A group studying crab populations counted the settlement of crab larvae every day for 15 months, including a day when a hurricane hit!

SMITHSONIAN ENVIRONMENTAL RESEARCH CENTER

www.serc.si.edu

On the shores of Chesapeake Bay, the Smithsonian Environmental Research Center (SERC) and its

The Smithsonian's National Museum of Natural History dedicates itself to studying the natural world and humans' place within it.

scientists seek answers to a wide range of questions related to coastal ecosystems. One major focus is the effect of human-induced environmental changes, which include pollution, climate change, and the spread of nonnative species. Because these changes are occurring in concert, and on land and as well as in marine waters, researchers are investigating how change in one area of an ecosystem affects the other elements of that ecosystem. In the longest-running experiment of its type, SERC has been monitoring the effect of increased carbon dioxide on marsh plants in special field chambers since 1987.

Manatee swimming

SMITHSONIAN TROPICAL RESEARCH INSTITUTE

www.stri.org

Some three million years ago, the Isthmus of Panama rose from the sea, forming a barrier between the Atlantic and Pacific oceans. This event had profound effects on Earth's climate and the evolutionary and ecological trajectory of marine communities in the region. With field stations on both sides of the isthmus, the

Smithsonian Tropical Research Institute (STRI) is the perfect place for investigating the differences between the two oceans and the changes that have occurred. One recent change documented by STRI scientists was the almost complete loss of formerly abundant long-spined sea urchins throughout the western Atlantic due to disease. The disappearance of the urchins, which are important grazers, has led to many reefs becoming overgrown by algae.

CARRIE BOW CAY MARINE FIELD STATION, BELIZE

www.nmnh.si.edu/iz/ccre.htm

Scientists at Carrie Bow Cay study everything from microbes to manatees in the reefs, sea-grass beds, and mangrove forests that surround the station. On the reef, researchers have observed a shift from one type of coral to another following an epidemic of the white band coral disease, and also studied the role of sea urchins, water flow, and sunlight on reef growth and health. Other researchers have manipulated nutrient flow through mangrove communities to better understand how this flow affects these important ecosystems.

Long-spined sea urchin

INDEX

ACKNOWLEDGMENTS & PICTURE CREDITS

The author would like to acknowledge the fine and supportive team of editors at Hylas Publishing. I would like to thank Molly Jacobs and Eva Dusek for expert input on corals; the 2005 Friday Harbor Laboratories Invertebrate Zoology students and TAs and my co-instructor Louise Page for tolerating my mad writing schedule; the Pfaff family for letting me work in their intertidal, where some of the photos in this book were taken; Molly Morrison for being a fun and supportive editor; and Daniel Froehlich, for providing bird expertise, helpful insights, and lots of support during the writing process.

The publisher wishes to thank consultant Paul Sieswerda, Aquarium Curator New York Aquarium; Brian Huber, Smithsonian National Museum of Natural History; Ellen Nanney, Senior Brand Manager with Smithsonian Business Ventures; Katie Mann and Carolyn Gleason with Smithsonian Business Ventures; Collins Reference executive editor Donna Sanzone, editor Lisa Hacken, and editorial assistant Stephanie Meyers; Hydra Publishing president Sean Moore, publishing director Karen Prince, editorial director Aaron Murray, art director Brian MacMullen, editor Molly Morrison, designers Erika Lubowicki, Ken Crossland, Eunho Lee, Pleum Chenaphun, Gus Yoo, and La Tricia Watford, editors Marcel Brousseau, Ward Calhoun, Suzanne Lander, Rachael Lanicci, Michael Smith, Liz Mechem, and Amber Rose; picture researcher Ben DeWalt; copy editor Glenn Novak; and indexer Cynthia Crippen; Wendy Glassmire of the National Geographic Society; Harriet Mendlowitz of Photo Researchers, Inc; and Kim Fulton-Bennett of the Monterey Bay Aquarium Research Institute.

The following abbreviations are used: PR–Photo Researchers, Inc.; SPL–Science Photo Library; JI–© 2006 Jupiterimages Corporation; SS–Shutterstock; IO–Index Open; IS–iStockphoto.com; BS–Big Stock Photos; NOAA–National Oceanic and Atmospheric Association; OAR–Oceanic and Atmospheric Program; NURP–National Undersea Research Program; NESDIS–National Environmental Satellite, Data, and Information Service; USFWS–U.S. Fish and Wildlife Service; USGS–United States Geologic Survey; NSF–National Science Foundation; NASA–National Aeronautics and Space Administration; GSFC–Goddard Space Flight Center; SI–Smithsonian Institute; AP–Associated Press; LoC–Library of Congress; NGIC–National Geographic Image Collection; NWPA–The North Wales Photographic Association; ACOUS–Arctic Climate Observations Using Underwater Sound; FS–Fotosearch; COML–Census of Marine Life; GI–Getty Images; WI–Wikimedia

(t=top; b=bottom; l=left; r=right; c=center)

Introduction: Welcome to Ocean Science
V NOAA VI NOAA 1 NOAA 1 NOAA 2 NOAA 3 NOAA 3 NOAA 3 NOAA

Chapter 1: The Vast Unknown
4 NGIC/Raul Touzon 5t JI 5b JI 6tl JI 6bl JI 7t NOAA Census of Marine Life/Rudd Hopcroft 7b SS/Kerry L. Werry 8tl JI 8b SS/Dennis Sabo 9tl PR/ Photo Researchers Inc 9tr SI/COML/Michael Vecchione 10tl NOAA /COML/Russ Hopcroft 10b NGIC/James P. Blair 11tl JI 11br Alamy/Jeff Rotman 12tl SPL/Alexis Rosenfeld 12cr SS/Peter Baxter 12br NOAA/NESDIS/ Office of Research & Applications 13t NOAA/COML/ Bodil Bluhm 14tl JI 14bl JI 15t NGIC/Michael Nichols 15br NOAA

Chapter 2: The History of Oceans
16 NGIC/Carsten Peter 17t SPL/Steve Munsinger 17b SI/NOAA 18tl JI 18br SPL/Chris Butler 19tl NGIC/O. Louis Mazzatenta 19cr SI/Alfred Harrell 20tl SPL/Sinclair Stammers 20bl SS/Mark Bond 21tl JI 21cr PR/Mark Garlick 22tl NWPA 22cr NWPA 23bl LOC 23tr Alamy 24tl SI/Eric Long 24cr Alamy/Mary Evans Picture Library 24bl SI/Donald Hulbert 25t Alamy/Mary Evans Picture Library 26tl NWPA 26crNGIC/Bruce Sale 26bl SPL 27tr SPL/Sheila Terry 28tl NOAA/Steve Nicklas 28tr LOC 28bl SI/Adam Booth 29tr NGIC/Marc Moritsch 30tl NOAA/OAR/NURP/Steve Nicklas 30bc NOAA 31tr NASA 31bl NASA

Chapter 3: Ever-changing Earth
32 JI 33t SI 33b SPL/Photo Researchers 34tl Fotosearch 34cr PR/Gary Hincks 34br PR/Zephyr Photo Researchers, Inc. 35tl IS/Mary Lane 36tl PR/W.Haxby, Lamont-Doherty Earth Observatory 36br PR/Mark Garlick 37tl SPL/Photo Researchers Inc. 37bl PR/Jon Lomberg 37tr FS 38tl PR/Gregory G. Dimijian, MD 38tr

PR/Daniel Sambraus 38br NOAA/NURP/OAR 39tr NASA/GSFC SeaWiFs Project/Orbimage 40tl NASA/ GSFC/ASTER Science Team 40tr PR/Philippe Psaila 40br NURP 41tr GI 41cr PR/DOE/Science Source 42bl NOAA/OAR/NURP/ University of Hawaii 42br PR/Dr. Ken MacDonald 43tr NASA 43cr PR/WorldSat International 44tl JI 44bc USGS 45tc NOAA/A. Malahoff/OAR/NURP/Universiy of Hawaii

Chapter 4: The Miraculous Molecule
46 NGIC/Paul Nicklen 47t JI 47b SPL/Francoise Sauze 48tl JI 48bl SPL/Bernhard Edmaier 49tr NGIC/Gordon Wiltsie 49br JI 50tl JI 50bl SPL/Tony McConnell 50br SPL/Alfred Pasikea 51br JI 52tl SS/Steffen Foerster Photography 52tr Alamy/Dennis Kunkel 52br NGIC/ Paul Nicklen 53tr SS/Dennis Sabo 54tl SPL/Bill Bachman 55tr NOAA/Chris Doley 56tl JI 56bl SPL/Alexis Rosenfeld 54tl JI 54bl SS/Gerard Dieckmann, K. Heumann 57tr SS/Danilo Ducak 57br ACOUS

Chapter 5: The Air Above
58 NASA 59t JI 59b NASA 60tl NASA Kennedy Space Center 60bl SPL/Andrew Syred 61c NASA/GSFC Scientific Visualization Studio 62tl SPL/Sinclair Stammers 62cr PR/Mark Garlick 62br SPL 63c PR/Dee Breger 64tl NOAA Ship Collection 64tr NASA 65r © Dr. Steven Hare, Int'l Pacific Halibut Comm. 65br NOAA Fisheries Collection 66tl NOAA 66br NASA 67br NASA 68tl NASA/NASA Goddard Space Flight Center 68bl NASA 69tl NASA/NGFC 69tl NWPA 70tl JI 70tr NASA 70br NASA 71tr JI 72tl NOAA 72tr NASA 72bl NOAA 73c NASA

Chapter 6: Water on the Move
74 NASA 75t NASA/Earth Observatory 75b IS 76t NASA 76b NGIC 77t NGIC 77c NOAA/Jamie Hall 78 NOAA 79tl NOAA/Commander John Bortniak 79br NASA/Earth Observatory 80bl NASA 80br NASA 81 NOAA 82tl NOAA/Steve Nicklas 83 NASA 84tl NASA/ Earth Observatory 85tr SPL NOAA/Richard Behn, NOAA corps 86tl NSF 86bl SPL/Los Alamos National Laboratory 87t NOAA 87b NSF 88tl NOAA 88b NOAA/Michael Van Woert 89tr IS/Jamie Wilson 89br NOAA and the Oceanic Museum of Monaco

Ready Reference
90cb LOC 90r LOC 91l LOC 92 93 WI/NASA 99tl SPL/Eye of Science 99cl SPL/M.I Walker 99cl SPL/Eye of Science 99bl SPL/Alfred Pasikea 99tr SPL/B.Murton/ South Hampton Oceanography Center 99cr SPL/Alfred Pasikea 99cr PR/Dr. Kari Lounatmaa 99br SPL/Dr. M. Rohde 100tl SPL/Science Source 100cl SPL/Astrid & Hanns-Frieder Michler100cl SPL/Andrew Syred100bl SPL/Jan Hinsch 100tc SPL/Sinclair Stammers 100c SPL/Alexis Rosenfeld 100c SPL/Michael Abbey 100c SPL/Sinclair Stammers 100bc SPL/Gregory Ochoki 100tr SPL/Juergen Berger 100cr SPL/Bob Gibbons 100br SPL/Bob Gibbons 101tl NOAA 101cl NOAA 101tl NOAA/NURP/OAR 101bl NOAA 101tc SPL/David Scharf 101c NOAA 101bc NOAA 101br SPL/Sinclair Stammers 101tr NOAA 101cr SPL/Nancy Sefton 101br NOAA 102bl NOAA 102c NOAA 102cr NOAA 103bl SS/J.McPhail 103r NOAA 103bar NOAA 104l NOAA 104r NOAA/OAR/NURP 105 NOAA

Chapter 7: Waves and Tides
106 NGIC/Cotton Coulson 107tl SS/ Pieter Janssen 107bl IS/Blache Designs 108tl IS/Ian MacDonnell 109tl IS 109tr WI 110tl JI 110br NGIC 11t NOAA/Sean Linehan/NGS 11lb NOAA 112tl NOAA 112b NASA/ Earth Observatory 113 NOAA 114tl SS/Tyler Olson 114b NASA/Earth Observatory 115 NIX 116tl NOAA/ Sean Linehan 116b NOAA 117tl NOAA 117tr NOAA 118tl IS/PicsofMaine.com 118bl Alamy 119tl PR/Georg Gerster 119tr NOAA/Captain Albert E. Theberge 120tl JI 120bl PR/Martin Bond 121t NOAA

Chapter 8: At the Ocean's Edge
122 NGIC 123tl NOAA/Commander John Bortniak 123bl NOAA/Anthony Piccolo 124tl NOAA/William Folsom 124cr IS/Stan Fairgrieve 124br NOAA/Dr. James P. McVey 125 NIX 126tl NOAA/Captain Albert E. Theberge 126b NOAA/David Sinson 127IS/Sherwood Imagery 128tl NOAA Courtesy of NPS-Canaveral National Seashore 128b NOAA/David Sinson 129tl NOAA/Fernando Arraya 129tr Washington State Department of Ecology 130tl NOAA/commander John Bortniak 130tr NGIC/Sisse Brimberg 130br IS 131tr NOAA/Mary Hollinger 132tl NOAA/Richard B Mieremet 132bl NOAA 132br NOAA/OAR/NURP and the University of North Carolina at Wilmington 133 JI 134tl NOAA 134br NOAA 135t NOAA/Personnel of NOAA Rainier 135b NOAA/Rear Admiral Harley D. Nygren 136tl NOAA 136bl NOAA/Richard B. Mieremet

137tr NOAA/William Folsom 137br NOAA/Dr. Gary E. Eddey

Chapter 9: Life Between the Tides
138 JI 139tl JI 139bl JI 140tl JI 140tr SS/Natthawat Wongrant 141tr SS/Dwight Smith 141bl NOAA/Bob Williams 142tl NOAA 142b Woods Hole Oceanographic Institute/ Jesus Pineda 144tl IO/FogStock, LLC 144cr PR/James Zipp 145tl NOAA 145br PR/Russ Curtis 146tl NOAA/Rick Crawford 146bl JI 147tr NOAA 147bl SS/Christina Tisi-Kramer 148tl IO/AbleStock 148br NOAA 149tr NGIC/Joel Sartore 149cr SS/Martin Bowker 150tl SS/Ian Bracegirdle 150bl NGIC/Nick Caloyianis 151tl SS/Theresa Martinez 151tr SS/Vladmir Ivanov 152tl NOAA/NESDIS/P.R. Hoar 152bl NOAA/ NOS/ Dr. Terry McTigue 152bc NOAA/Jack Terrill 153 SS/Vera Bogaerts

Chapter 10: Life in the Open Sea
154 NOAA/Justin Marshall 155tl JI 155bl SS/Justin Kim 156tl SS/Andrei Volkovets 156br NOAA 157tc SPL/Claire Ting 157tr NOAA 158tl PR/Eric V. Grave 158b NASA/Jacques DesCloitres and MODIS Land Rapid Response Team 159tr SPL/Volker Steger 159bl PR/Dante Felonio 160tl NASA/SeaWiFS Project 160cr SS/Christoffer Vika 161 NOAA 162tl PR/Alexis Rosenfeld 162bl PR/Alexis Rosenfeld 163tr PR/Science Pictures Limited 163bl PR/F. Stuart Westmorland 164tl SPL/Andrew G. Wood 164br NOAA 165tr NOAA 165bl PR/George C. Lower 166tl NOAA/Dade W. Thornton 166bl NOAA 167tl NOAA/Lieutenant Philip Hall 167tr NOAA National Marine Fisheries Service 168tl SS/Steffen Foerester 168tr SS/Ian Scott 169tr SS/Bateleur 169bl NOAA/Quartermaster Joseph Schebal

Chapter 11: Life in the Deep
170 NGIC/Emory Kristof 171t NOAA Mountains in the Sea Research Team 171b NOAA 172tl NOAA/M. Youngbluth OAR/NURP Harbor Branch Oceanographic Institution 172tr NOAA 172bc NOAA 173 NOAA/HBOI/Brooke et al 174tl NOAA/Jason Chaytor 174tr NOAA/OAR/NURP/E. Williams 175tr NOAA/Mountains in the Sea Research Group/the IFE Crew 175bl NOAA/Mountains in the Sea Research Group/the IFE Crew 176tl SPL/Dee Breger 176bl MBARI 177t MBARI 177bSPL/Andy Harmer 178tl NOAA/ OAR/NURP/I. MacDonald 178tr NOAA/OAR/ NURP/ Texas A&M University 179tr NOAA 179bl NOAA 180tl NOAA/OAR/NURP Harbor Branch Oceanographic Institution/M. Youngbluth 180bl NOAA 180br PR/Dr. Paul A. Zahl 181 NSF/MSU-CBE c 2002 Angela Bowlds MSU Bozeman Bioglyphs Project 182tl SPL/Institute of Oceanographic Sciences/NERC 182br SPL/Alexis Rosenfeld 183bl NOAA 183tr NSF/Scripps Institution of Oceanography

Chapter 12: The Future of Oceans: Threats and Solutions
184 U.S. Fish and Wildlife 185t U.S. Fish and Wildlife 185b NOAA/Maria Brown 186tl SS/Keith Levit 186bl SPL/Alexis Rosenfeld 187bl SPL/Geospace 187tr SPL/Jerry Mason 188tl NOAA/Mary Hollinger 188tr NOAA/William Folsom 189tl NOAA/Top-E Pescola; Center-M.Deflorio; Greenpeace 189bl SS/Vixique 190tl NOAA 190bl SPL/Pat & Tom Leeson 191tr NOAA 191tl NOAA Sea Grant Program/Dr. James P. McVey 192tl SS/Tracy Carolyn Lee 192tr Regulatory Fish Enc. 192br Regulatory Fish Enc. 193bl U.S. Fish and Wildlife 193tr SS/Stephen Snyder 194tl SPL/David Nunuk 194bl NGIC/Cotton Coulson 195t NASA/Visible Earth 195bl NOAA 196tl NOAA 196c US Fish and Wildlife 196bl NOAA/Yuri A. Zuyev and the University of St. Petersburg 197 Courtesy of the Washington State Department of Fish and Wildlife/Russell Rogers 198tl NOAA/NESDIS/ORA/ and Michael Van Woert 198bl NSF/Zina Deretsky 199tl NGIC/Norbert Rosing 199bl NOAA/NDOC/Mary Hollinger 199tr SPL/Fred McConnaughey 200tl LOC 200bl NOAA/Glenn Allen 201bl NOAA/P.R. Nelson 201tr NOAA/Gulf of Farallones National Marine Sanctuary 202tl NOAA Office of NOAA Corps Operations 202l NOAA/Justin Marshall 203tl U.S. Fish and Wildlife 203br NOAA/ Captain Robert A. Pawlowski

At the Smithsonian
210b SI/ Dane A. Penland 211bl NOAA 211tr NOAA

Cover
Front Gregory Ochocki/Photo Researchers, Inc.
Background Photo Researchers, Inc.